研究生“十四五”规划精品系列教材

Quality Graduate Teaching Materials for the 14th Five-Year Plan of Xi'an Jiaotong University

International Journal Article Writing and Conference Presentation (Humanities and Social Science)

国际期刊论文写作与会议交流（文科类）

主　编　王　华　王伊蕾
副主编　杨　蕾
编　者（以姓氏笔画为序）
田荣昌　杨文健　邹郝晶
罗　玲　姜冬蕾　董　清

图书在版编目（CIP）数据

国际期刊论文写作与会议交流. 文科类 / 王华, 王伊蕾主编; 杨蕾副主编. -- 西安 : 西安交通大学出版社, 2024. 12. -- ISBN 978-7-5693-3875-1

Ⅰ. G312；G321.5

中国国家版本馆CIP数据核字第2024DF4464号

国际期刊论文写作与会议交流（文科类）

International Journal Article Writing and Conference Presentation (Humanities and Social Science)

主　　编　王　华　王伊蕾
副 主 编　杨　蕾
责任编辑　庞钧颖
数字编辑　宋庆庆
责任校对　牛瑞鑫
装帧设计　伍　胜

出版发行　西安交通大学出版社
（西安市兴庆南路1号　邮政编码710048）
网　　址　http://www.xjtupress.com
电　　话　（029）82668357　82667874（市场营销中心）
（029）82668315（总编办）
传　　真　（029）82668280
印　　刷　陕西奇彩印务有限责任公司

开　　本　787 mm × 1092 mm　1/16　**印张** 12.75　**字数** 300千字
版次印次　2024年12月第1版　2024年12月第1次印刷
书　　号　ISBN 978-7-5693-3875-1
定　　价　49.00元

如发现印装质量问题，请与本社市场营销中心联系。
订购热线：（029）82665248　（029）82667874
投稿热线：（029）82668531　（029）82665371

前言

Preface

国际学术期刊和国际学术会议通常采用英语作为工作语言，各国学者通过在国际学术期刊上发表英文学术论文实现相互交流、学习和推广科技成果的目的。国际学术期刊论文发表和国际学术会议展示是各国科研工作者的一项重要工作。研究生作为科技队伍的新生力量，应具备学术英语写作、发表与汇报展示的能力，这些能力是其科研素质的重要组成部分，也是高校培养国际型、创新型人才的重要衡量指标之一。近年来国内研究生学术写作和会议交流的教学活动取得了长足的发展，已形成以体裁分析法为理论指导的主流教学模式，据此编写的教材也应运而生。然而，目前的写作教学与教材较多关注不同学科学术论文的共性，较少关注学科间的差异，教材中的例文选择更倾向于较易被理解的社会性或公共性话题，不能有效地突出不同学科学术论文的语篇特点。实际上，不同学科的学术论文存在较大差异，只强调共性而忽略特性无法有效对学生进行指导，甚至可能产生误导。目前，虽然已有一些针对特定学科领域的学术论文写作教材，但数量有限，仍无法满足市场需求。“国际学术期刊论文写作与会议交流”系列教材正是在这样的背景下编写而成的。本套教材包括理工、医学和文科类三个分册，教材基于三大领域内的真实学术期刊论文和国际学术会议资料，充分展现了三大类学科论文的结构和语言特点，基于功能教学法、语类分析、跨文化交际、语域分析和脚手架等理论框架，对各领域学生有针对性地进行写作指导。本套教材既可供高校教师、研究生（博士生、硕士生）、高年级本科生及相关科研人员学习使用，也可用于对国际学术会议参会人员的培训。

本册为文科类分册，依据学术写作教学的经典理论——体裁教学理论，将解构、建模、再认、模仿共建和独立建构的教学思想融入教材编写，实现教学方法从以教师为主体向以教师为主导、学生为主体、产出为目的的转变，搭建脚手架，以用促学，学以致用。选用的主要素材为近年来发表在 SSCI 和 AHCI 一区或二区学术期刊上的论文，涵盖教育学、语言学、新闻传播学、心理学、经济金融等领域的研究。本册按照学术期刊论文的通用结构，即 IMRD（Introduction-Methods-Results-Discussion）安排教材内容，对论文结构及对应句法、时态、典型用语等语言特征进行分析并设置相应的写作练习。

本册教材共十个单元。第一单元和第二单元涉及学术论文写作的准备阶段，旨在让学生了解其专业领域内有影响力的学术期刊、文献的检索方法和引用方式、研究性论文的体裁特征等，培养学生的修辞意识。第三单元至第七单元主要介绍实证研究性论文主要组成部分的写作策略，包括引言、研究方法、研究结果、讨论与结论、标题与摘要等内容，侧重分析各部分的信息要素及典型语言特征。第八单元关注论文的写后修改。第九单元帮助学生了解学术期刊论文发表的流程和环节，涉及论文作者与学术期刊编辑及论文审稿人之间书信往来的措辞、语言修改与润色等问题。第十单元的主题是国际会议交流，主要介绍口头汇报和海报展示等会议展示形式，从视觉呈现和口头呈现两个方面对会议报告者进行指导。

本教材的每个单元都提供了大量的实例，并配以相关练习供教师和学生使用。在使用本教材时，建议教师可以进行下述操作：

1）在每个自然班内，可将学生按照专业最接近原则建组，每组 3 至 4 人。

2）要求每位学生从自己专业的国际一流学术期刊中挑选三篇影响因子较高的论文作为个人的语料积累。论文必须是实证研究性论文，内容与自己的研究兴趣有关。

3）师生在课堂上共同分析教材例文的结构、信息要素和典型的语言特征。学生课后仔细研读和分析自己所选的三篇本专业学术期刊论文，确认论文的结构和信息要素是否和教材所述的典型结构一致，如有差异，分析产生差异的原因，同时关注这些信息要素中的典型语言特征。

4）学生比较自己所选择的三篇论文在结构和语言方面的异同，加深对语言功能的理解。学生还可以在同一小组内对比不同学术期刊论文的研究范式和写作常规，分析不同学术期刊论文的共性和差异，并以小组为单位在课堂上进行汇报和交流。

在本册教材编写的过程中，我们参阅并借鉴了大量国内外相关文献资料和同类教材，在此向所有相关作者表示深深的感谢！如需获取本书相关教学资料，请发送邮件至 yilei.wang@xjtu.edu.cn。此外，我们还咨询了西安交通大学张晓鹏教授，宾夕法尼亚州立大学陆小飞教授，南加州大学丹尼尔 · 西尔弗（Daniel Silver）教授，加利福尼亚大学塔马拉 · 塔特（Tamara Tate）教授、马克 · 沃沙尔（Mark Warschauer）教授，以及香港理工大学冯德正教授，感谢各位专家帮助我们更加深入地了解二语习得、教育学、新闻传播学等领域的学术研究。西安交通大学出版社的编辑们也为本书的出版付出了大量的汗水和辛勤的劳动，在此一并向他们致以诚挚的谢意！本书虽经反复讨论和精心编写，但由于编者才疏学浅，加之时间仓促，书中不妥之处和谬误在所难免，恳请学界各位专家、学者及广大读者朋友提出宝贵的批评、修改意见和建议。

编　者

二〇二四年六月

目　录

Table of contents

Unit 1

Preparing for Your Writing

Learning objectives

In this unit, you will

- understand research writing; and
- learn how to manage your sources.

Self-evaluation

Before embarking on your research writing journey, it is crucial to assess your current strengths and areas for improvement. This self-evaluation will help you understand your readiness for the challenges ahead and guide you in honing the necessary skills for successful academic writing. Consider the following key areas:

- **Research skills assessment:** Evaluate your ability to conduct thorough research, gather relevant sources, and critically analyze information.
- **Writing proficiency:** Assess your writing skills, including organization, clarity, coherence, and adherence to academic conventions.
- **Time management:** Consider your time management skills and how effectively you can allocate time for research, writing, editing, and revising.
- **Citation and referencing:** Reflect on your understanding of citation styles (e.g., APA, MLA, Chicago) and your ability to properly cite sources.
- **Critical thinking:** Evaluate your capacity for critical analysis, argumentation, and the integration of diverse perspectives in your writing.

Research writing is a detailed process that involves investigating a topic, analyzing information, and presenting findings in a structured manner. It is commonly used in academic settings but is also applicable in many professional fields. Before conducting research writing, we need to do a lot of preparation works. Preparing our writing, whether it is for academic, professional, or personal purpose, involves several important steps to ensure clarity, coherence, and impact. Here is a guide to help you prepare your writing effectively.

Understanding and being able to analyze **rhetorical situations** can contribute to well-organized and audience-focused academic writing. The term "rhetorical situation," being widely used by professional writers, refers to any set of circumstances that involves at least one person using some sort of communication to modify the perspective of at least one other person. But many people are unfamiliar with the concept of "rhetoric." For some people, "rhetoric" may imply speech that is simply persuasive. For others, "rhetoric" may imply something more negative like "trickery" or even "lying." In this case, to appreciate the benefits of understanding what rhetorical situations are, we must first have a more complete understanding of what rhetoric itself is. In brief, "rhetoric" is any communication used to modify the perspectives of others. But this is a very broad definition that calls for more explanations.

With its origin traced back to ancient Greece, "rhetoric" is defined as primarily an awareness of the language choices we make. Rhetoric also refers to the persuasive qualities of language. The ancient Greek philosopher Aristotle strongly influenced how people have traditionally viewed rhetoric. Aristotle defined rhetoric as "an ability, in each particular case, to see the available means of persuasion." Since then, Aristotle's definition of rhetoric has been reduced in many situations to mean simply "persuasion." At its best, this simplification of rhetoric has led to a long tradition of people associating rhetoric with politicians, lawyers, or other occupations noted for persuasive speaking. At its worst, the simplification of rhetoric has led people to assume that rhetoric is merely something that manipulative people use to get what they want (usually regardless of moral or ethical concerns).

However, over the last century or so, the academic definition and use of "rhetoric" has evolved to include any situation in which people consciously communicate with each other. In brief, individual people tend to perceive and understand just about everything differently from one another (this difference varies to a lesser or greater degree depending on the situation, of course). This expanded perception has led a number of more contemporary rhetorical philosophers to suggest that rhetoric deals with more than just persuasion; it should be regarded as a set of methods people use to identify with each other—to encourage each other to understand things from one another's perspectives. From interpersonal relationships to international peace treaties, the capacity to understand or modify another's perspective is one of the most vital abilities that humans have. Hence, understanding rhetoric in terms of "identification" helps us better communicate and evaluate all such situations. The parts of the rhetorical situation are as follows.

Audience: Your audience is person or group of people you are writing for. For example, in psychology courses, this is often your professor or teaching assistant, although you might also be asked to write for a "general audience of psychologists" or to your classmates. Your instructor may or may not indicate who your audience is for your paper, so it is always good to ask. In articles, it is more complex—a combination of reviewers, journal editors, and readers in your area of interest. Your audience's expectations about your writing determine

- formatting and style;
- tone of the piece;
- the amount of technical language or jargon used; and
- the amount of information you assume the audience have already known.

Audience expectations aren't always straightforward. For example, if you are taking a course in psycholinguistics and you are writing a critical review of research on semantic priming, your primary audience for the course is your instructor. While your instructor may know what "semantic priming" is, you may still be expected to define this concept in your paper so that your instructor knows that you know what it is. Part of the instructor's expectation in this case is that you can clearly define key psycholinguistic concepts discussed in class in your term paper.

Purpose: While the overall purpose of your term paper or experimental report may be clear (i.e. either to pass the course, or to convey the results of your research), more specific purposes for writing your report are not always so. When you are prewriting and drafting, ask yourself not only what your larger purpose is, but also what additional purposes you may have.

Context: The context is the larger writing situation in which you find yourself. Are you writing for a class? Are you writing an internal report to your advisor? Are you writing an article for submission to a journal? The context in which you are writing is another important factor that helps you determine the style, format, and content of your piece.

Setting: Lastly, all rhetorical situations occur in specific settings or contexts or environments. The specific constraints that affect a setting include the time of author and audience, the place of author and audience, and the community or conversation in which authors and/or audiences engage.

Time: In the above description, it refers to specific moments in history. It is fairly common knowledge that different people communicate differently depending on the time in which they live. Americans in the 1950s, overall, communicate differently than Americans in the 2000s. Not that they necessarily speak a different language, but these two groups of people have different assumptions about the world and how to communicate based on the era in which they live. Different moments in time can be closer together and still affect the ways that people communicate. Certainly, scientists discussed physics somewhat differently the year after Einstein published his theory of relativity than they did the year before Einstein's publication. Also, an author and audience may be located at different times in relation to one another. Today, we appreciate Shakespeare's *Hamlet* a bit differently than people who watched it when it first premiered four hundred years ago. A lot of cultural norms have changed since then.

Place: Similarly, the specific places of authors and their audiences affect the ways that texts are made and received. At a rally, the place may be the steps of a national monument. In an academic conference or lecture hall or court case, the place is a specific room. In other rhetorical situations, the place may be the pages of an academic journal in which different authors respond to one another in essay form. And, as mentioned about authors' and audiences' backgrounds, the places from which audiences and authors emerge affect the ways that different texts are made and received.

Conversation: In various rhetorical situations, conversation can be used to refer to the specific kinds of social interactions among authors and audiences. Outside of speaking about rhetorical situations, conversation usually means specific groups of people united by location and proximity like a neighborhood; conversation usually refers to fairly intimate occasions of discussion among a small number of people. But in regard to rhetorical situations, both of these terms can have much larger meanings. In any given rhetorical situation, conversation can refer to the people specifically involved in the act of communication. For instance, consider Pablo Picasso who used cubism to challenge international notions of art at the time he painted. Picasso was involved in a worldwide conversation of artists, art critics, and other appreciators of art many of whom were actively engaged in an extended conversation with differing assumptions about what art is and ought to be. Sometimes, authors and audiences participate in the same conversation, but in many instances, authors may communicate in one conversation (again, think of Shakespeare four hundred years ago in England) while audiences may participate in a different community and conversation (think of scholars today in any other country in the world who discuss and debate the nature of Shakespeare's plays). The specific nature of authors' communities and conversations affect the ways that texts are made while the specific nature of audiences' conversations affect the ways that texts are received and appreciated.

UNDERSTANDING RHETORICAL SITUATIONS

Below is a simplified abstract of a sample article. Read the provided text and answer the following questions to explore the rhetorical situations.

History, modernity, and city branding in China: a multimodal critical discourse analysis of Xi'an's promotional videos on social media

Abstract

In the digital age, cities around the world are mobilizing various symbolic resources to rebrand their images through social media. Against this background, this study investigates how Xi'an, a second-tier developing city in China, constructs its digitalized urban imaginary using the popular social media platform of TikTok. A semiotic framework is developed to model Xi'an's urban imaginary as evaluative attributes and to elucidate how they are constructed through linguistic and visual resources in short videos on TikTok. The analysis of 294 videos shows that Xi'an highlights its dual identity as a modern metropolis and a historical city. The modern metropolis image is characterized by the personification of Xi'an as a stylish, young, popular, and international microcelebrity; the historical city image is constructed through recreating the Great Tang Dynasty and revitalizing local folk art. The characteristics of city branding discourse reflect China's *wanghong* economy, urban policies and the affordances of social media.

Questions:

1. What is the intended audience of the research paper?
2. How are the language and content tailored to this audience?
3. What is the main purpose of the study as outlined in the excerpt?
4. What is the genre of the research paper and its conventions?
5. What is the author's stance on Xi'an's digitalized urban imaginary?

Rhetorical analysis framework

In academic article analysis, you need to focus on how the following elements interact to serve the article's persuasive purpose. Consider how effectively the article addresses its audience, how well it establishes its argument within the existing research landscape, and the potential impact of its findings on its intended audience and broader society. You may use the following rhetorical analysis framework to analyze the academic article. For example,

Title: *The impact of urban green spaces on mental health: A case study*

Abstract: *This article examines the relationship between urban green spaces and mental health outcomes among residents of metropolitan areas. Through a mixed-methods approach combining quantitative data analysis and qualitative interviews, the study aims to contribute to the growing body of evidence supporting urban planning policies that prioritize green space development. The findings suggest that increased access to green spaces significantly correlates with improved mental health metrics, including reduced symptoms of stress and anxiety.*

Reference: *Wu, Yishan. 2022. The impact of urban green spaces on mental health: A case study [J]. Second Language Education.*

When analyzing this hypothetical academic article, consider the following aspects of the rhetorical situation.

Purpose: Understand the article's goal, which is to influence urban planning and public health policy by demonstrating the positive impact of green spaces on mental health.

Audience: The intended audience includes academics, researchers, urban planners, and policymakers. The language, style, and presentation of data are tailored to these groups, presumed to have a specific interest and competence in the subject matter.

Speaker/Writer: The authors position themselves as experts in urban planning and public health through their methodology and analysis. Their credibility is established through their systematic approach to the research question.

Context: The study is situated within a broader concern for mental health in urban environments, aiming to influence policy and planning towards healthier cities.

Message: The article's message is conveyed through a structured presentation of evidence and arguments supporting the role of green spaces in improving urban mental health.

Medium: Published in an academic journal, the article uses formal academic language, visual data representations, and references to previous research to communicate its findings and analysis.

Task 1.1 Please choose an appropriate research paper to make presentations and discuss with your groupmates.

1. Preparation.

Schedule presentations: Participants are allocated specific time slots to present their works. Ensure there's enough time for the presentation and a feedback session afterward (Time: 5–10 minutes for presentation and 5 minutes for feedback. Format: e.g., PowerPoint, speech, demonstration. Key points to cover: e.g., purpose, audience, message, medium.).

2. Presentation.

(1) Presenting work: Participants present their work, focusing on how they addressed their assigned rhetorical situation. Participants may use of visual aids, handouts, or digital presentations to support their message.

(2) Audience engagement: Encourage the audience to take notes and think of questions or feedback to provide after the presentation. If appropriate, the presenter can include interactive elements or questions to engage the audience.

3. Feedback.

(1) Structured feedback: After each presentation, open feedback from the audience. Use a structured format, such as compliment, suggest, question, to guide the feedback process.

(2) Written feedback: Optionally, participants and the facilitator can provide written feedback using a standardized form. This allows for more detailed and private feedback that presenters can reflect on later.

4. Reflection and discussion.

(1) Group reflection: After all presentations, facilitate a group discussion. Encourage participants to share what they learned from the exercise, both from presenting and from observing their peers.

(2) Discussing challenges: Have an open discussion about any challenges encountered during the exercise and explore solutions or strategies to overcome these challenges in future communications.

(3) Highlighting key learnings: Summarize key takeaways from the exercise, emphasizing the importance of tailoring communication to the rhetorical situation, the value of constructive feedback, and the role of practice in improving communication skills.

Task 1.2 Write your plan based on the above presentation.

1. Reflection assignment: Participants are required to reflect on the feedback they received and write a plan for how they will incorporate this feedback into their future work.

2. Re-presentation opportunity: Participants can present revised versions of their work, demonstrating how they've applied the feedback.

MANAGING YOUR SOURCES

Please search journals online or in your university library. Pick three journals in your field and look at the articles in the journal. Then compare in a group what you and your group members have found.

Journal name	Majors of the journal	Impact factors	Article genre published in the journal	Time to publication	Article length published in the journal

Selecting a target journal

How do the readers select a target journal?

Selecting the proper journal for your manuscript will provide the chance of getting published easily and quickly. You'd better think about the journal you want to publish in advance and should make a choice by the time you begin to write your paper. The choice of the journal determines a size of the audience accessing and using your work and the professional reputation and prestige which may flow from the publication. Your right choice of the journal will help you to optimize the speed and ease of publication, accrue your professional reputation, and gain the access to your desired audience. These factors are interwoven and it can help to develop a publication plan to maximize your publication success. One of the first considerations is whether the journal peer reviews the articles that it publishes. The peer-review process is important for establishing the quality of your work, and you should seek peer-reviewed journals to publish in if you wish to develop a research profile. Of course, the journal of your choice may not choose to accept your article and you are advised to have a list of preferred journals to turn to if you are rejected from your first choice.

Who may read the papers in the journal?

The audience for a journal is largely determined by the scope and aims of the journal, the journal's reputation and history of publishing in the field, and the accessibility of the journal to researchers (e.g. is it expensive? Does it have Open Access options for authors? Is it published by a small publisher with limited distribution?). Internet access to journal titles, abstracts, and homepages has allowed many more journals to be accessible to a wider audience. However, some users may not wish to pay for access to a paper, so journals that are widely bought by institutions will have a wider audience for practical purposes. New journals may also take time to develop an audience. Check the journal website and publisher to see whether a journal you are considering is widely distributed.

What is high-quality of the journal?

There is no easy way to assess the quality of a journal or the contribution of a journal to a research discipline over time. A number of indices have been developed to provide information on the relative speed and volume of citation to journals, and these indices can give some guidance about the relative popularity and usage of a journal. The most commonly used measure of journal impact is the journal impact factor.

The journal impact factor for a specific year represents the average frequency with which articles published in the journal during the preceding two years have been cited within that same year. This index provides a measure of the average recent use of articles in a given journal.

Other measures of the influence of a journal on its field of research are

- journal immediacy index, calculated as the number of citations to articles in the year with respect to the number of articles published in that year, giving a measure of how rapidly the average article in a given journal is used; and
- journal cited half-life, calculated as the number of publication years from the current year that account for 50% of citations received by the journal, giving a measure of the longevity of use of the average article in a given journal.

Which indices are used to evaluate the journal?

Statistics on citation number as a measure of journal quality should be used with an awareness of the purpose for which the statistics are gathered and the limitations of these indices. The indices described above all measure the rate or volume of citation of the average article in a journal. They are measures of the journal and not the individual articles. The number of citations for your article can also be calculated and may be higher or lower than the average for the journal. For example, getting your articles read, cited, or used is about reaching the right audience.

Sometimes the right audience may not be the readership of the journal with the highest impact factor. Other things to consider when assessing indices for ranking journals are listed as followings:

- Comparing journals from different fields of research may not be meaningful (e.g. mathematics researchers cite very few journals, whereas papers in molecular biology journals cite dozens).
- The calculation of some indices is prone to inflate the relative contribution of journals which include sections for discussion and review (rather than original research).
- Citation-matching procedures are strongly affected by sloppy referencing, editorial research, characteristics of journals, some referencing conventions, language problems, author-identification problems, and unfamiliarity with names from some countries.
- Published indices are calculated from selected list of journals. This list largely excludes journals published in non-English-speaking-countries, and may not include new journals still establishing their reputation.

- Journal ranking based on indices can change over time. The impact factor for one journal increased, one decreased, and one remained relatively stable. However, articles in each of the three journals will continue to be cited on their individual merit.

The above elements can be helpful to develop a publication plan to maximize your publication success. Another important thing to consider is whether the journal peer reviews the articles that it publishes. The peer-review process is important for establishing the quality of your work, and you should seek peer-reviewed journals to publish in if you wish to develop a research profile. Of course, while submitting your article to your preferred journal, it is important to be aware that there is a possibility of rejection. It is prudent to prepare a list of alternative journals that align with your research area, which can serve as viable options if your initial choice does not accept your submission.

What is the aim and scope of the journal?

In the Introduction and Discussion sections, the journals that are most often cited will be most likely to accept work in your field. Examine some of the key articles in the field of your research, and check which journals are often cited. By following back through the literature, you should be able to develop a mind-map of the journals in the field of your research. Check the websites or issues of these journals to identify those with scope and aims that are most appropriate for your manuscript.

When do the readers publish the paper?

Journals want to publish submissions quickly to ensure they attract authors who are doing innovative and new work. You may also want to publish your research quickly to ensure that others do not publish similar work before you, and to increase your publication and citation record for promotions and grants. If time to publication is important to you, you should check journal websites or recent issues to see whether they report the average time to publication. Journals which publish an online version of the paper before the print version will usually have a faster time to publication.

How do the readers use or publish the papers?

Some journals charge fees for publishing manuscripts. Fees may be based on a fixed cost or on the number of pages, or they may be charged for publishing color illustrations or for reprints. Check whether the journal charges for any part of the publishing process before you submit your manuscript. You may also want your research to be accessible to a wide range of readers who do not have access to libraries or other subscriptions to journals in your field. Many journals now offer to provide Open Access to your paper (i.e. to make it accessible for free download without subscription to the journal) if you pay an upfront fee. If you intend to (or are obligated by your institution) pay for Open Access, it is advisable to verify whether the journal you have selected provides this service.

Which types of journal articles do the readers encounter?

Scholar publishes various types of scholarly articles which have different aims and requirements. Most of the articles published will be one of the following.

Original research: Research articles refer to novel and innovative analysis, or experiment which add to the present knowledge and information available on a particular topic. An original article emphasizes the complete description of advance research carried out including Introduction, Method, Results, Discussion, and Conclusion supported by strong statistical results along with their significance sections. Original articles are the crucial and most important type of papers used as primary sources.

Review articles: Review articles summarize information from the valuable published scientific literature. These types of articles compile significant studies, make comparisons, address limitations, and frequently identify specific gaps or issues, thus highlighting the need for future research. Though the review articles don't report any novel or original research, one can get an idea about the current state of a topic without reading all the published works in the field.

Case reports: Case reports discuss the unique, unusual, rare features of a disease, symptoms, signs, diagnosis, pathogenesis, new therapeutic approaches, an unexpected drug interaction, important scientific observations and follow-up of an individual patient including pictures that contribute to the existing knowledge in the field of medicine.

Case reports contain literature review of other cases providing essential information to physicians of all specialties in medical profession.

Short communications: High-impact, most interesting research outcomes often deserve to be shared quickly. Short communications are short, brief, or rapid communications that present novel and significant material for rapid distribution. They are concise format used to report improvements to existing methods, a new application, or a resource. These need to be reported quickly as the need to communicate their findings is very high.

Editorials: Editorials address diverse topics, relevant concerns, changes, science, politics, news regarding the journal and its editorial management, content or policies. They are generally written by the editorial team or the editorial board members.

Letters: Letters to the editor are intended for comments on articles published in the journal and must cite published references to support the writer's argument or constructive comments on the subjects of articles. A letter to the editor can be written by anyone who wants to share an opinion.

Opinion-based articles: Opinion-based articles typically take the form of essays, offering a personal perspective to evaluate overarching concepts or prevailing ideas within a specific field. Opinion articles present the author's viewpoint on the elucidation, analysis, or methods used in a particular study. It allows the author to comment on the strength and weakness of a theory or hypothesis. Opinion articles are usually based on constructive criticism and should be backed by evidence. Such articles promote discussion on current issues concerning science. These are also relatively short articles.

Corrigenda: Corrigenda correct errors and omissions in published papers. They should be as brief as possible, and describe no new methods or results.

Methodologies or methods articles: These articles present a new experimental method, test or procedure. The method described may either be completely new, or may offer a better version of an existing method. The article should describe a demonstrable advance on what is currently available.

Task 1.3 To optimize the outcomes from publishing your manuscript, it is a good way to develop a publishing strategy. An integral aspect of the publication strategy involves carefully selecting a preferred journal to submit your manuscript. In order to make this choice, select the three or four preferred journals in your field that you think would accept your manuscript. Then answer the following questions for each journal and record the answers.

1. Has the journal published similar work with a similar level of novelty to yours in the last 3 years? Record "yes" or "no" (if "no," think carefully before submitting your manuscript to this journal).

2. Does the journal's scope and the content of recent articles match the components of your manuscript, i.e. subject methods and results? (write down the main type of papers, e.g. plant physiology, non-molecular studies, etc.).

3. What is the measure of relative journal quality/impact which is most important to you and your field of research? Record the score or measure for each journal (e.g. journal impact factor or journal cited half-life).

4. What is the journal's time to publication (this may be on the journal's website or recorded for each article in the journal)? Record the time or a score for fast or slow (e.g. less than 3 months from acceptance =fast; more than one year =slow).

5. Does the journal have page charges or provide Open Access if you want it (and can you pay if payment is required)?Examine the journal scores you have recorded in the following table and rank the journals in order of overall preference, taking all criteria into consideration.

6. Rating preferred journals in terms of key criteria for maximizing your publication success.

Journal name	Recent publication of similar work and novelty	Match of scope and recent content to your work	Journal quality/ impact	Time to publication	Open Access costs

Locating the target journal

When you start your background research, one of the early steps is finding and reading the scientific literature related to your science project. Mentors are a great resource for recommendations about which scientific papers are critical for you to read and you should ask your mentor, or other experts in the field, for advice. But there'll also be times when your mentor is busy or isn't up-to-date on a particular experimental method, in which case, you'll need to be proactive and hunt for papers on your own. It has been observed that using regular search engines to simply input search terms is not particularly effective in yielding desired results.The chapter you get back will be a wide mixture of websites, and very few will be links to peer-reviewed scientific papers. To find scientific literature, the best thing to use is an academic search engine.

There are many different academic search engines. Some focus on a single discipline, while others have citations from multiple fields. There are a handful of free, publicly available academic search engines that can be accessed online; some of these are listed in Table 1.1. The remainder, like the ISI Web of Science, is subscription-based. Universities and colleges often subscribe to academic search engines. If you can not find what you need using a free search engine, you may be able to access these resources from computers in a university or college library. Consult the school's library webpage, or call the library directly, to find out to which academic search engines they subscribe to and whether or not you'd be allowed into the library to access them.

Table 1.1 Some free academic search engines

Academic Search Engine	URL	Disciplines
ScienceDirect	http://www.sciencedirect.com/science/search	All
Web of Science	www.webofscience.com	All
PubMed	www.ncbi.nlm.nih.gov/pubmed	Life science
IEEE Xplore	ieeexplore.ieee.org/Xplore/guesthome.jsp	Electronic and electrical engineering/ Computer science
National Agriculture Library	www.nal.usda.gov	Agriculture
Education Resources Information Center	eric.ed.gov	Education

Here are a few tips to help you get started with the academic search engines:

- Each search engine works slightly differently, so it is worth taking the time to read any available help pages to figure out the best way to use it.
- When you are beginning your literature search, try several different key words, both alone and in combination. Then, as you view the results, you can narrow down your focus and figure out which key words can best describe the kinds of papers in which you are interested.

As you read the literature, go back and try additional searches using the jargon and terms you learn while reading.

Note: The results of academic search engines come in the form of an abstract. You can read this abstract to determine if the paper is relevant to your science project. Additionally, you'll receive a full citation (author, journal title, volume, page numbers, year, etc.), allowing you to find a physical copy of the paper. Search engines do not necessarily contain the full text of the paper for you to read. A few, like PubMed, do provide links to free online versions of the paper, when one is available. Read on for help finding the full paper.

Task 1.4 Read the following citation and answer the questions.

> Gurnham, D. (2005). The Mysteries of Human Dignity and the Brave New World of Human Cloning. *Social and Legal Studies*, 14 (2), 197–214.

1. Is this article available in your library?

2. How can you get a copy of this article?

3. What type of periodical is this?

Search literature in Web of Science

Web of Science is the world's most popular publisher-independent citation database. Guided by the legacy of Dr. Eugene Garfield, inventor of the world's first citation index, Web of Science is the most powerful research engine, delivering best-in-class publications and citation data for confident discovery, access, and assessment.

The multidisciplinary platform connects regional, specialty, data and patent indexes to Web of Science Core Collection. The comprehensive platform allows you to track ideas across disciplines and time from almost 1.9 billion cited references from over 171 million records.

Over 9,000 leading academic, corporate and government institutions and millions of researchers trust Web of Science to produce high-quality research, gain insights, and make informed decisions that guide the future of their institution and research strategy. Below are the key steps in using Web of Science.

Step 1. Formulating your research questions

Research questions are often formulated broadly. Try to formulate your research question clearly. What do you want to know and how do you think you can get an answer to your question? Bear in mind your research question throughout the whole writing process (e.g., searching for literature, processing results and writing). Think critically about what you want to find out. For example:

Discuss current approaches to childhood obesity in the UK.

Step 2. Deciding your search elements

Divide your search questions into search aspects. You can use the Building Blocks method to tailor your search. The search query can also act as a good base for you: Keep asking yourself whether the search strategy you are using is going to provide an answer to your question. Make a note of the elements you are going to use. Think of as many English synonyms as possible for each element.

When using the Building Blocks method, you divide your question into aspects. This involves extracting the keywords that can be used to find articles. These aspects form the basic parts for your search strategy. Take, for example, the following research question: *Discuss current approaches to childhood obesity in the UK.* The key concepts in the question are:

Search aspect 1	Search aspect 2	Search aspect 3
approach	*childhood*	*obesity*

Decide which terms you could use for each of these aspects, and think of as many synonyms as possible.

Search aspect 1	Search aspect 2	Search aspect 3
approach	*childhood*	*obesity*
treatment	*youth*	*overweight*
…	…	…

Step 3. Using databases

Web of Science is often the first choice of database for all discipline. It is made up of several different sections, including Science Citation Index, Social Science Citation Index, and Arts & Humanities Citation Index.

Step 4. Constructing a search strategy

Construct a search strategy to use Web of Science. When you find a strategy that is effective, translate the query to use it in other databases.

Boolean logic

"OR" will search for articles containing any of the terms we choose. Use "OR" to combine synonyms, alterative spellings or related items. "AND" will search for articles which contain all of the terms we have chosen.

We can expand those keywords into collections of synonyms. You may want to broaden your search to include plurals, grammatical variations and spelling variations, so you can use "TRUNCATION" or "WILDCARDS."

Truncation / Wildcards

- The asterisk (*) represents any group of characters, including no character (e.g. "s*food" will find *seafood* and *soyfood*).

- The question mark (?) represents any single character (e.g. "wom?n" will find *women* and *woman*)
- The dollar sign ($) represents zero or one character (e.g. "isch$emia" will find *ischaemia* and *ischemia*). The dollar sign can be placed in the middle or at the end of the word.

You can also use combinations of these wildcard tools to get the broadest possible variation: e.g. "organi?ation*" will find *organisation*, *organization*, *organisations*, *organizations*, *organizational*, and *organisational*.

Recognize the key phrases in your search—this will help you improve the relevance of your search results: searching for hormone replacement therapy might retrieve papers which use all the words, but not necessarily in this phrase.

Phrase searching: using double quotes

To search for an exact phrase, enter it in quotes, e.g. "endometrial cancer."

Our strategy for the search now looks like this:

Search aspect 1	Search aspect 2	Search aspect 3
approach	*childhood*	*obesity*
treatment	*youth*	*overweight*
*Interven**	*child**	*"BMI"*

You may be alarmed at the number of hits you get for this first layer of your search. Don't worry—once all the terms are combined, the number of hits you have to look through will be much more realistic.

Step 5. Evaluating your search

You usually have to carry out more than one search, as the initial search often generates either too few or too many results. You may also find that too many of the articles are not relevant to your topic. Return to your research question and make the necessary changes to your search using the tips. You can click on "search history" to view all the lines of your search results.

Task 1.5 **Formulate your research question and use search engines to find out your results. Consider the steps described above.**

1. Fill in: Research question.

2. Decide the search elements.

Search aspect 1:

Search aspect 2:

Search aspect 3:

3. Use databases, e.g. Web of Science Core Collection.

4. Construct a search strategy.

Aspect 1:

Aspect 2:

Aspect 3:

5. Evaluate the search again, with the changes you have made. Copy and paste your final search here. Are you satisfied with the results?

Reference managing software

What are citation tools?

Citation tools help you store, organize, and share your research citations. They also automatically format your bibliographies and in-text citations into whatever style you need (e.g. APA, MLA, Chicago, and many more).

Which tool should I use?

There are four citation managers that are very trusted, that is, EndNote Basic, Mendeley, RefWorks, and Zotero. These citation managers provide similar basic features—they allow you to save citations, organize them into folders or libraries, and generate bibliographies and citations as you write. Once you choose a tool, know that you can always change your mind; sources can easily be transferred between tools. (Please note that attachments don't always transfer.)

The best way to figure it out is to set up an account on one of the tools and try it! It can also help to make an individualized appointment with a graduate student librarian at your library to learn more about choosing and using the tools.

Task 1.6 **Citation tools help you save, organize and share your references. They also let you automatically format in-text citations and bibliographies with the click button. Please list *pros* and *cons* of four citation tools, and tell your partners which one is right for you.**

Task 1.7 **Please choose your references to direct export from databases (or import text files from databases/add references manually), then share the screenshots of these references with your teachers and partners.**

Unit task

Academic Source Searching Skills

Imagine you are a graduate student conducting research on the impact of mindfulness-based interventions on reducing anxiety symptoms in adolescents. You need to find academic sources to support your literature review on this topic (participants apply your knowledge of search techniques and database navigation relevant to your academic discipline). Do the following to finish the task.

Step 1: Formulate a search query that effectively captures the key concepts of your research topic. Consider including synonyms or related terms to broaden your search.

Step 2: Use the provided computer lab and access the PsycInfo database, which specializes in psychology and behavioral sciences literature.

Step 3: Conduct a search using your formulated query and explore the search results.

Step 4: Refine your search results by applying filters such as publication date (e.g., last 5 years), document type (e.g., peer-reviewed articles), and methodology (e.g., empirical studies).

Step 5: Select two relevant articles from the search results that align with your research topic and objectives.

Step 6: Evaluate the credibility and relevance of the selected articles based on criteria such as author credentials, publication venue, research methodology, and theoretical framework.

Step 7: Discuss within your group any challenges encountered during the search process and share insights on effective search strategies and database navigation.

Materials Needed:

- Access to a computer lab with Internet connectivity
- Projector or screen for demonstrations
- Handouts or slides summarizing key concepts and search strategies
- Sample search scenarios or exercises
- Evaluation forms for feedback

Unit 2

Understanding Research Article Structure

Learning objectives

In this unit, you will

- understand research article structure and peer review process; and
- learn common language features of academic writing.

Self-evaluation

Read the following questions and give your understanding.

- How many elements do research articles include?
- How are the research articles organized?
- What are the language features of academic writing?

The structure of a research article is designed to systematically present research findings and their significance within the scientific community. While there are variations across disciplines, most research articles adhere to a widely accepted format known as IMRaD, which stands for Introduction, Methods, Results, and Discussion. Below is a detailed breakdown of the typical sections found in a research article, following the IMRaD structure, along with other common components.

Title is a concise statement that conveys the focus of the research. It should be specific enough to give a clear idea of the study's nature and scope.

Abstract is a succinct summary of the entire study, including its purpose, methodology, main findings, and conclusions. It allows readers to quickly grasp the essence of the article.

Introduction section sets the stage for the study, outlining the background, the research problem, and its significance. It also states the objectives and hypotheses of the research, establishing the context and justifying the study.

Literature Review section (sometimes being integrated into the Introduction section) reviews existing research on the topic, highlighting gaps that the current study aims to address. It situates the research within the broader academic conversation.

Methods (Methodology) section describes the research design, materials, participants, and procedures in detail, enabling replication. It also explains the analytical techniques used to examine the data.

Results section presents the findings of the study without interpretation, using text, tables, and figures for clarity. This section details the data collected and the outcomes of the statistical analyses or other methods of analysis.

Discussion section interprets the results, linking them back to the research questions and the existing literature. It explores the implications, significance, and potential limitations of the findings, and may suggest areas for future research.

Conclusion section summarizes the key findings and their relevance, reinforcing the importance of the study and its contributions to the field.

The research is presented in a logical, accessible manner, facilitating peer review and engagement by the wider academic community.

Task 2.1 Read an academic article related to your research and complete the following tasks.

1. Title: Write down the title of the article and explain how it reflects the theme of the research.

2. Abstract: Summarize the content of the abstract, including the research purpose, methods, results, and conclusions.

3. Introduction: Identify the research problem and background information presented in the Introduction section. List the research objectives and questions.

4. Literature Review: Summarize the main points of the Literature Review. Identify any research gaps the authors highlight.

5. Methods (Methodology): Describe the research design (qualitative, quantitative, or mixed methods).List the participants, sampling methods, and data collection techniques.

6. Results: Summarize the key findings of the study. Mention any tables, graphs, or quotes from participants that were used.

7. Discussion: Interpret the results in the context of the research questions. Compare the findings with previous studies and discuss their implications for educators.

8. Conclusion: Summarize the main findings and discuss the significance of the research and potential future directions.

9. References: List five sources cited in the article and check if the citation format is correct.

LANGUAGE FEATURES OF ACADEMIC WRITING

Examine the following texts and identify any significant features. What kind of text does the extract come from and how does the language differ between the texts?

Text 1

> **Introduction to Pitch**
>
> **2/1 Pitch names and notation**
>
> Playing any note on a piano produces a fixed sound. The sound gradually fades away, but it does not go up or down. Music is made up from fixed sounds such as this.
>
> Many instruments (including all the stringed instruments and the trombone) are capable of producing an infinite number of fixed sounds between any two notes on a keyboard, with only minute differences between them. It is the same with the human voice. But in practice all instruments and singing voices too, normally use only the particular notes of the keyboard. When a player such as a violinist "tunes" his instrument, he is trying to find exactly the one fixed sound he wants. All the other notes in the music will be placed in relation to this one note.
>
> If one note is played on the keyboard and then another note is played anywhere to the right

of it, the sound of the second note is said to be higher than that of the first. A note to the left of it would produce a lower sound. In the same way men's voices are said to be lower than those of women or young boys. The technical word referring to the height or depth of sound is pitch.

On the keyboard, groups of two black notes alternate with groups of three black notes. This makes it easy to distinguish between the white notes, which are given the letter names from A to G. A is always between the second and third of the group of three black notes. After G comes A again.

Text 2

The following lines are from Shakespeare's *Romeo and Juliet*, Act 2, Scene 2, during the famous balcony scene.

***Juliet*:**

Hist! Romeo, hist! O, for a falconer's voice
To lure this tassel-gentle back again!
Bondage is hoarse, and may not speak aloud;
Else would I tear the cave where Echo lies,
And make her airy tongue more hoarse than mine
With repetition of my Romeo's name.

In this scene, Juliet wishes she could call Romeo back like a falconer recalling a bird of prey (a "tassel-gentle" refers to a male falcon). She laments the restrictions ("bondage") that prevent her from calling loudly after him.

Text 3

This paper examines interaction in written text through the interplay between the notions of text averral and attribution (Sinclair, 1988). Text averral is evidenced in the unmarked parts of the text, where the utterances are assumed to be attributed to the author. Attribution, the counterpart of text averral, is the marked case where the sources of authority are clearly signaled.

It is hoped that this study will add to our knowledge about the characteristics of different types of text, and illuminate the way for students who find themselves lost amidst the echoes of the multiple voices they hear within the same text.

Text averral and attribution are basic notions for the organization of interaction in written text. The assumption is made that the author of a non-fictional artefact (Sinclair, 1986) avers every statement in his or her text so long as he/she does not attribute these statements to another source—whether that source is other or self. Averral is manifested in various ways in the text—negatively, through absence of attribution, and positively, through commenting, evaluating or metastructuring of the discourse. Attribution, on the other hand, is signaled in the text by a number of devices of which reporting is an obvious one.

Text 1

____

____

____

Text 2

____

____

____

Text 3

____

____

____

Complexity

Written language is relatively more complex than spoken language. Written texts are lexically dense compared to spoken language—they have proportionately more lexical words than grammatical words. Written texts are shorter and have longer, more complex words and phrases. They have more noun-based phrases, more nominalizations, and more lexical variation.

Written language is grammatically more complex than spoken language. It has more subordinate clauses, more "that/to" complement clauses, more long sequences of prepositional phrases, more attributive adjectives, and more passives than spoken language.

The following features are common in academic written texts:

- Noun-based phrases
- Subordinate clauses/embedding
- Complement clauses
- Sequences of prepositional phrases
- Participles
- Passive verbs
- Lexical density
- Lexical complexity
- Nominalization
- Attributive adjectives
- Adjectival groups as complements

Halliday compares a sentence from a spoken text:

"You can control the trains this way and if you do that you can be quite sure that they'll be able to run more safely and more quickly than they would otherwise, no matter how bad the weather gets."

With a typical written variant:

"The use of this method of control unquestionably leads to safer and faster train running in the most adverse weather conditions."

The main difference is the grammar, not the vocabulary.

Other equivalents are given below:

Spoken	Written
Whenever I'd visited there before, I'd ended up feeling that it would be futile if I tried to do anything more.	*Every previous visit had left me with a sense of the futility of further action on my part.*
The cities in Switzerland had once been peaceful, but they changed when people became violent.	*Violence changed the face of once peaceful Swiss cities.*
Because the technology has improved its less risky than it used to be when you install them at the same time, and it doesn't cost so much either.	*Improvements in technology have reduced the risks and high costs associated with simultaneous installation.*
The people in the colony rejoiced when it was promised that things would change in this way.	*Opinion in the colony greeted the promised change with enthusiasm.*

Task 2.2 Rewrite the following sentences in a more typically written style.

1. Because the jobs are even more complex, programmes to train people will take longer.

2. I handed my essay in late because my kids got sick.

Formality

In general, "formality" means in an essay that you should avoid

- colloquial words and expressions, e.g. *stuff*, *a lot of*, *thing*, *sort of*, etc.;

- abbreviated forms, e.g. *can't*, *doesn't*, *shouldn't*, etc.;
- two-word verbs, e.g. *put off*, *bring up*, etc.;
- sub-headings, numbering and bullet-points in formal essays—but use them in reports; and
- asking questions.

Below are some examples.

Colloquial words and expressions

With women especially, there is a lot of social pressure to conform to a certain physical shape.

With women especially, there is a great deal of social pressure to conform to a certain physical shape.

Significantly, even at this late date, Lautrec was considered a bit conservative by his peers.

Significantly, even at this late date, Lautrec was considered somewhat conservative by his peers.

Abbreviations

The radical restructuring of British politics after 1931 doesn't lie in the events of 13–28 August, but in the changing attitudes within the National Government.

The radical restructuring of British politics after 1931 lies not in the events of 13–28 August, but in the changing attitudes within the National Government.

Two-word verbs

A primary education system was set up throughout Ireland as early as 1831.

A primary education system was established throughout Ireland as early as 1831.

This will cut down the amount of drug required and so the cost of treatment.

This will reduce the amount of drug required and so the cost of treatment.

(There is often a choice in English between a two word verb and a single verb. For example, bring up/raise, set up/establish.)

Formal/Informal

Formal	**Informal**
Finally	*In the end*
Immediately	*At once*
Therefore	*So*

Task 2.3 Rewrite the following sentences, replacing the informal expressions with a formal equivalent.

1. It focused on a subject that a lot of the bourgeois and upper-class exhibition-going public regarded as anti-social and anti-establishment.

2. Later Florey got together with Paul Fildes in an experimental study of the use of curare to relieve the intractable muscular spasms which occur in fully developed infection with tetanus or lockjaw.

3. When a patient is admitted to a psychiatric inpatient unit, the clinical team should avoid the temptation to start specific treatments immediately.

4. The first National Government wasn't intended to be a coalition government in the normal sense of the term.

5. These aren't at all original or exotic but are based on the ordinary things that most people tend to eat.

6. Dieters often feel that they should totally get rid of high-fat and high-sugar foods.

7. Thus when a Gallic bishop in 576 converted the local Jewish community to Christianity, those who turned down baptism were expelled from the city.

8. Western scholars gradually turned out a corpus of translations from the Arabic and studies of Islam.

Accuracy

In academic writing you need to be accurate in your use of vocabulary. Do not confuse, for example, "phonetics" with "phonology," or "grammar" with "syntax."

When using English for academic purposes, it is important to be accurate both in speaking and writing. It is, however, very difficult to produce language which is intelligent, appropriate and accurate at the same time. It is therefore important to break down the task into stages: an ideas stage and an accuracy stage.

Speaking

When you are speaking, you need to prepare well. Do not prepare only your ideas; prepare your language as well. Michael Wallace's advice is very useful. Write out your spoken presentation in the way that you intend to say it. This means that if you are working from a piece of your own written work, you must do some of the work of writing the paper again. Written language is different from spoken language. Your seminar presentation will probably take less time than the written paper it is based on and you cannot summarize while you are standing at the front of the room. When you have written out your talk, you will need to carefully check it for accuracy. Another possibility is to record your presentation and watch it later. While you are watching it, look out for mistakes—a useful way to do this is to transcribe sections of your talk, as it is easier to notice mistakes when they are written down. You can then try to give the talk again.

Writing

When you are writing, you need to rewrite and edit your work carefully. In his seminal work *Writing and the Writer* (1982), Frank Smith distinguishes between "composition" and "transcription" in writing. "Composition" is deciding what you want to say, and "transcription" is what you have to do to say it. He give the following advice:

"The rule is simple: Composition and transcription must be separated, and transcription must come last. It is asking too much of anyone, and especially of students trying to improve all aspects of their writing ability, to expect that they can concern themselves with polished transcription at the same time that they are trying to compose. The effort to concentrate on spelling, handwriting, and punctuation at the same time that one is struggling with ideas and their expression not only interferes with composition but creates the least favorable situation in which to develop transcription skills as well."

So whether you are speaking or writing, it is important to have a specific time for working on accuracy.

Task 2.4 Many people would say there was a "mistake" in each of the following sentences. Can you identify them? For each sentence, please do four things:

- **Mark the faulty word or words.**
- **Briefly describe what is wrong.**
- **Try to write a correct or improved version.**
- **Decide how important the mistake is.**

1. A rentcharge is the right to receive an annual sum out of the income of land every year.
2. All of the solutions considered so far have only involved Legendre functions of even order.
3. Before spelling out exactly what this means, it is worth first asking whether translating machinery is necessarily irreversible.
4. Being in charge, the accusation was particularly annoying to me.
5. Finally, it enables the therapist to assess the effectiveness of his own clinical skills and, hopefully, to improve them in future.
6. He is a former student who I've not seen for years.
7. He is an unskilled labourer and works at odd jobs, but he don't do any of them very well.
8. However, there were other patients whose lives had ended by suicide.
9. I have now discussed the proposals for replacing all the computers with my colleagues.
10. It is felt that less people would be put off attending with other problems if the stigma of going to a clinic could be dispensed with.

Hedging language

Academic writing, particularly in the realm of scientific literature, is frequently regarded as a medium dedicated to presenting factual content and disseminating information. However, it is now recognized that an important feature of academic writing is the concept of cautious language, often called "hedging" or "vague language." In other words, it is necessary to make decisions about your stance on a particular subject, or the strength of the claims you are making. Different subjects prefer to do this in different ways. Below are some examples of the language used in hedging.

Introductory verbs

seem, tend, look like, appear to be, think, believe, doubt, be sure, indicate, suggest

Certain lexical verbs

believe, assume, suggest

Certain modal verbs

will, must, would, may, might, could

Adverbs of frequency

often, sometimes, usually

Modal adverbs

certainly, definitely, clearly, probably, possibly, perhaps, conceivably

Modal adjectives

certain, definite, clear, probable, possible

Modal nouns

assumption, possibility, probability

That-clauses

It could be the case that...

It might be suggested that...

There is every hope that...

To-clause + adjective

It may be possible to obtain...

It is important to develop...

It is useful to study...

Compare the following:

It may be said that the commitment to some of the social and economic concepts was less strong than it is now.

The commitment to some of the social and economic concepts was less strong than it is now.

The lives they chose may seem overly ascetic and self-denying to most women today.

The lives they chose seem overly ascetic and self-denying to most women today.

Weismann suggested that animals become old because, if they did not, there could be no successive replacement of individuals and hence no evolution.

Weismann proved that animals become old because, if they did not, there could be no successive replacement of individuals and hence no evolution.

Task 2.5 Identify the hedging expressions in the following sentences.

1. There is no difficulty in explaining how a structure, such as an eye or a feather, contributes to survival and reproduction; the difficulty lies in thinking of a series of steps by which it could have arisen.

2. For example, it is possible to see that in January this person weighed 60.8 kg for eight days.

3. For example, it may be necessary for the spider to leave the branch on which it is standing, climb up the stem, and walk out along another branch.

4. Escherichia coli, when found in conjunction with urethritis, often indicate infection higher in the uro-genital tract.

5. There is experimental work to show that a week or ten days may not be long enough and a fortnight to three weeks is probably the best theoretical period.

6. Conceivably, different forms, changing at different rates and showing contrasting combinations of characteristics, were present in different areas.

7. One possibility is that generalized latent inhibition is likely to be weaker than that produced by pre-exposure to the CS itself and thus is more likely to be susceptible to the effect of the long interval.

8. For our present purpose, it is useful to distinguish two kinds of chemical reaction, according to whether the reaction releases energy or requires it.

9. It appears to establish three categories: the first contains wordings generally agreed to be acceptable, the second wordings which appear to have been at some time problematic but are now acceptable, and the third wordings which remain inadmissible.

Unit task

Proof-Reading Written English

When writing English for academic purposes, it is important to be accurate. It is, however, very difficult to produce language which is intelligent, appropriate and accurate at the same time. It is therefore important to break down the task into an ideas stage and an accuracy stage. In the accuracy stage, all your ideas are on the paper and you can concentrate on accuracy. You can carefully read your work and correct your mistakes. This is proof-reading—and it takes time.

Furthermore, in the same way that it is difficult to concentrate on ideas and accuracy at the same time; it is difficult to check your work for all kinds of mistakes at the same time. You therefore need to check your work several times, for different purposes. For example, you could first check your verbs, then check your prepositions and articles etc.

Proof-read the following text: how many mistakes can you find?

Comparative study of animal help to show how man's space require are influenced in his environment. In animals we can observing the direction, the rate, and the extent of changes of behaviour that follow changes in space available to them as we can never hope to do in men. For one thing, by using animals it am possible to acelerate time, since animal generations is relatively short. Scientist can, at forty years, observe four hundred forty generations of mice, while has in the same span of time seen only two generations of his own kind. And, off course, he can be more detatched about the fate of animal.

Unit 3

Introducing Your Study

Learning objectives

In this unit, you will

- know the function of the Introduction section;
- learn the factors should be involved in writing the Introduction section;
- command the linguistic strategies for writing an effective Introduction section; and
- think about what your own Introduction section will look like.

Self-evaluation

Scan the QR Code, read the Introduction section of the Sample Articles provided and answer the following questions.

- How do the authors start the Introduction sections?
- How do the authors introduce the topics?
- How do the authors end the Introduction sections?

Think it over:

What kind of information should be covered in the Introduction section?

The section of Introduction is the first part of your paper, which acts as a blueprint of sorts for the paper as a whole, illustrating the central case behind your writing and the line of reasoning that you will follow throughout. The paper's Introduction section should begin with a general overview of the subject, followed by a precise statement that explains your ultimate intent and how it will be built upon in the body.

It is a key section for the quadripartite: authors, journal editors, reviewers and readers. As your primary audience, journal editors are likely to scan here for evidence to answer the following questions: Is it suitable for publication in the journal? Is the contribution significant?

As your secondary audience, peer reviewers will definitely start reading from this section to find out whether the work contribute something new, and whether this part provide the basic logic of whole paper.

For readers, an effective paper's Introduction section possesses the ability to captivate readers, generating a strong interest that compels them to continue reading the entirety of the paper. The Introduction section is a perfect place for the reader to set the scene and make a good first impression. It serves as a transition by moving the reader from the world outside of the paper to the world within.

Although the Introduction section comes at the start of your paper, it doesn't have to be the first thing you write—in fact, it's often the very last part to be completed (along with the abstract). It's a good idea, though, to write a rough draft of your Introduction section at the beginning. If you write a paper proposal, you can use this as a template, as it contains many of the same elements. However, you should revise your Introduction section throughout the writing process and return to it at the end, making sure it matches the content of your paper.

As an old saying goes "a good beginning is half done," it is essential to draw readers in with a strong beginning and set the stage for your research with a clear focus. To some extent, a carefully crafted Introduction section acts as a springboard, establishing the order and direction for the entire paper. Readers will be more inclined to read the paper if the Introduction section is clear-cut, organized, and engaging. Therefore, an effective Introduction section is of vital importance to a research paper.

INFORMATION CONVENTION

Below is the Introduction section of Sample Article 5 (SA5). Read it and fill in the table with the numbers of sentences that perform the given functions.

Factors Hindering Student Participation in English-Speaking Classes: Student and Lecturer Perceptions

Introduction

1 **1** In the Vietnamese context, the English subject is compulsorily taught in numerous educational institutions, from primary to tertiary levels, since it plays an essential role in the development of the country. **2** Therefore, educational institutions are expected to enhance the effectiveness of English teaching and learning. **3** One of the solutions is to propose some programs, often named "Advanced programs," "High-quality programs in English," and "Mainstream programs" to promote English language abilities. **4** The current study was employed in a tertiary institution in the Mekong region with Advanced and High-quality programs. **5** These programs involve English as the medium of instruction (EMI), requiring non-English major (NEM) students to advance their English proficiency in reading, writing, or presenting reports and materials in English for academic demands and professional careers in the future. **6** Students are taught in General English courses to acquire adequate linguistic competence (in pronunciation and lexicogrammar) and the target language skills (listening, speaking, reading, and writing) to master the contents of their major courses through EMI. **7** Furthermore, by the time of graduation, Vietnamese NEM learners are expected to reach level 4 of the Vietnam Foreign Language Framework (VFLF), which is equivalent to level B2 in the Common European Framework of References for Languages (CEFR) (Phuong, 2017). **8** Consequently, the call to acquire English, especially speaking skills, has progressed dramatically at the tertiary level.

2 **9** Notwithstanding the consequences of achieving communicative competence in learning English as a foreign language (EFL), Vietnamese students' English proficiency levels remain unsatisfactory (H. T. Nguyen, 2018). **10** This is especially the case of verbal competence, which is below expectations when students have finished tertiary education (T. T. Nguyen & Nguyen, 2016). **11** One reason for this distance "far from the expectation" is students' low participation in English-speaking classrooms (Vo et al., 2018). **12** Many reasons have been found as the factors inhibiting Vietnamese students' engagement in English-speaking classes, such as their traits (obedience, shyness, timidness, and the likes), teacher-student power distance, and so on (Aubrey et al., 2022; H. T. Nguyen et al., 2014; Tran, 2013a, 2013b). **13** However, Vietnamese teachers and students do not have enough opportunities to share their viewpoints inside or outside the classrooms. **14** Hence, it remains necessary to investigate what inhibits students' classroom participation throughout the teaching and learning process from the perspectives of lecturers and students in English-speaking classrooms. **15** Consequently, this study aims to check whether there is any significant difference between lecturer and student perceptions of the factors hindering EFL student participation at a tertiary institution.

The function of the sentence	Sentence number
providing general background information	
providing a brief overview of previous studies in the general area	
beginning to narrow down the research scope	
describing a gap in the research	
describing the paper itself	

Overall structure of the Introduction section

Although what is covered in the Introduction section may vary from one discipline to another and from one journal to another, some basic issues are similar, which need to be answered before conducting the research:

- What does the reader need to know to understand the paper?
- What specific aspect(s) of the topic will you address?
- Why is this research worth doing?
- What did you aim to find out and how did you approach it?
- What are the innovative points that you will cover?
- What are your research questions?

Some basic elements could be summarized from the above questions. These information elements (IEs) can be classified into three information chunks: (1) establishing the context and introducing the topic; (2) identifying research gaps; and (3) stating your research questions. Some of IEs are optional, and some of IEs are obligatory.

Briefly, the author should offer enough information to guide readers to the current research and engage readers in understanding the rest of the paper. In the Introduction section, authors usually start from a general subject area and move on to a particular field of research. Therefore, the overall structure of the section can be thought of as a funnel shape (as shown in Figure 3.1)—the top presenting the broadest view and the rest narrowing down to the specific research problem. Such a structure responds to two kinds of competition: competition for readers and competition for research space.

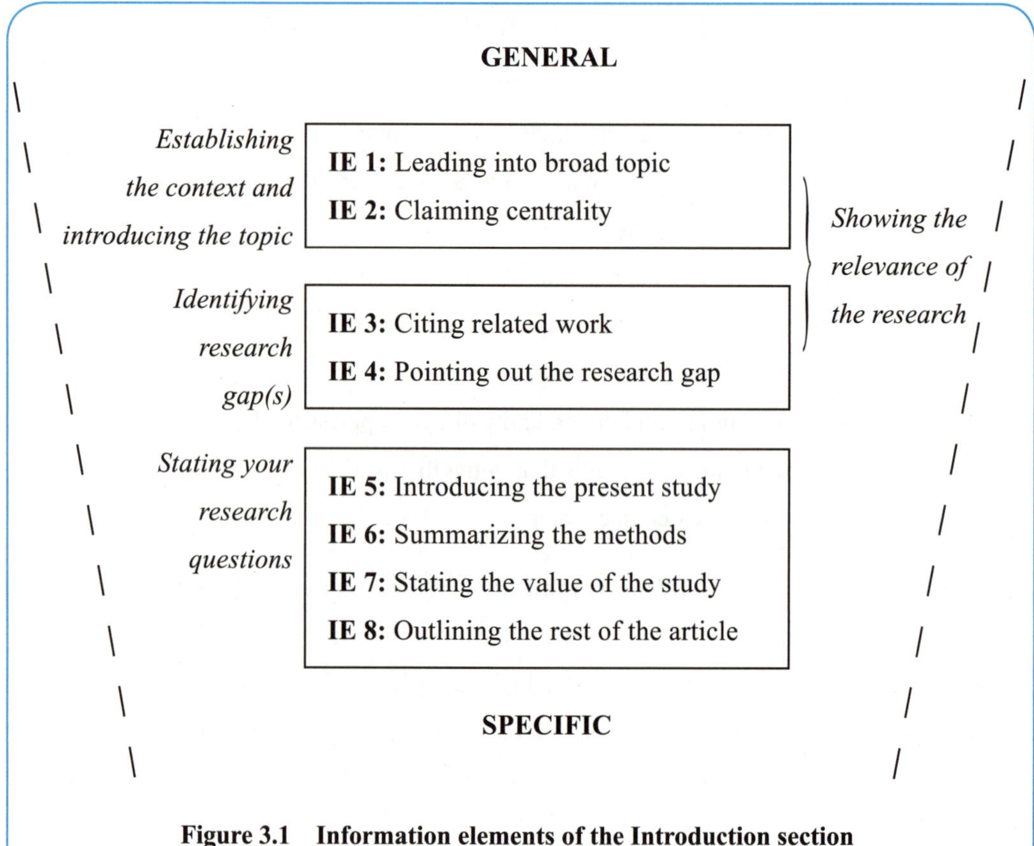

Figure 3.1 Information elements of the Introduction section

Note: Not all the information elements may appear in a paper. There are disciplinary variations. The elements may not always follow a given order. Some may appear in cycles. Some may be integrated and not clear-cut from each other.

Task 3.1 Read the Introduction section in SA5 again and answer the questions below.

1. How is the information organized from general to specific? Does the structure of the Introduction section present a funnel shape?

2. Is it a good idea to omit Sentence 1? Why?

3. Why does the author include references in Sentence 2?

Task 3.2 **Complete the table and take down notes of the words or expressions used to signal the information elements (IEs).**

IEs	Words or expressions
Leading into broad topic	
Claiming centrality	
Pointing out the research gap	
Introducing the present study	

Establishing the context and introducing the topic

The essence of writing a good paper lies in the skillful application of captivating storytelling, where an engaging opening assumes utmost significance, which depending on offering possible and general background information. Possible necessary general background information should be given in this step so as to make your paper an attractive opening.

In most cases, authors should establish a context to help readers understand how the research fits into a wider field of study by using a few sentences or several paragraphs in the Introduction section. The Introduction section tends to begin with a problem of wide interest to claim topic centrality, or to appeal to the discourse community. Authors often begin by leading into a broad topic which aims to spark interest and show why this is a timely or important topic for a paper (for example, by mentioning a relevant news item, debate, or practical problem) that would generally be accepted as fact by the members of their target discourse community.

Task 3.3 **Below is part of the Introduction section of Sample Article 1 (SA1). Read it and finish the True or False task.**

> **1** Linguistic complexity (LC) needs to be carefully considered when selecting or adapting second language (L2) teaching materials, given its potential effect on text comprehensibility and processing (e.g., Bailey & Heritage, 2014) and the important role of comprehensible input in L2 acquisition (Gass, 2015). **2** Numerous readability formulas for text difficulty (TD) classification exist, mostly based on TD judgment data from first language (L1) speakers (see Benjamin, 2012, for a review). **3** Such formulas may be unreliable for selecting or adapting L2 teaching materials, as L1 and L2 speakers likely perceive TD differently given their vastly different language learning experiences (Nahatame, 2021).

1. The topic of this article is Linguistic complexity. (　　)
2. Sentence 1 introduced the context of the paper. (　　)
3. The function of sentence 2 is to claim centrality and indicate the topic. (　　)
4. Sentence 3 identified the shortcomings in previous research. (　　)
5. This paragraph followed such logic: linguistic complexity–L2 acquisition–TD-formulas–L1 and L2 speakers. (　　)

Narrowing down your focus

After a brief introduction to your general area of interest, zoom in on the specific focus of your research. For example: What geographical area are you investigating? What time period does your research cover? What demographics or communities are you researching? What specific themes or aspects of the topic does your dissertation address? You should also clearly define the scope of your research, that is, the boundaries of what you will and won't cover.

In this stage, authors may cite relevant literature (as needed) to support the claims being made and connect the work to existing knowledge. Whether it is necessary to include references depends on the field and the topic of the paper. In addition, sentences written in the present perfect tense are also used in citing related work, expressing what has been found over an extended period in the past and up to the present. The present tense is frequently employed in research articles due to its function in English, wherein it conveys information perceived as universally true.

Narrowing down your focus can be unfolded in this way containing three levels of upper and lower scopes from the largest to the smallest relationships: first, create an "Earth" for your readers; then, isolate one "Country" within this earth; and finally, center your readers attention on a concrete "Individual" in the country (See Figure 3.2).

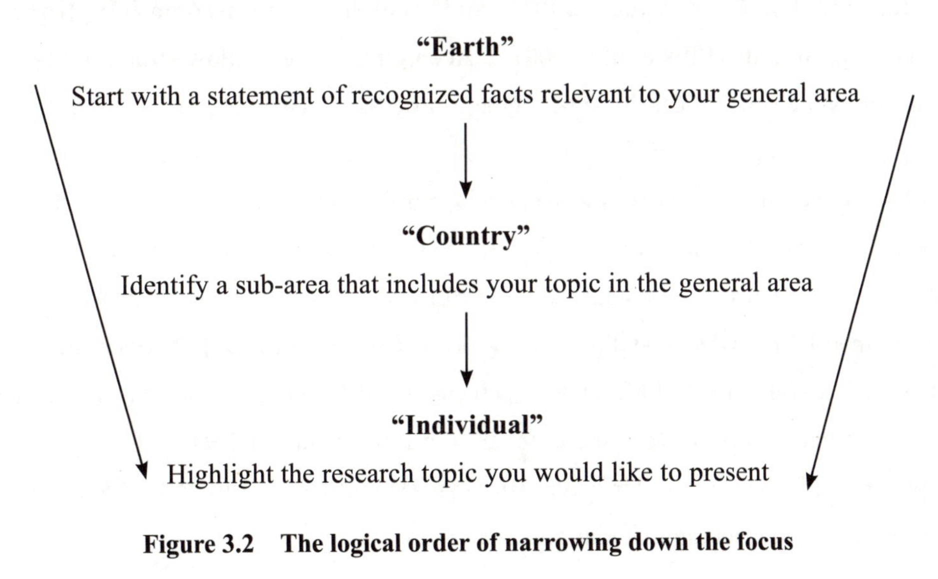

Figure 3.2 The logical order of narrowing down the focus

Task 3.4 Read the whole part of the Introduction section of SA1 and identify the "Earth," the "Country," and the "Individual." Complete the table under the text.

Testing the Relationship of Linguistic Complexity to Second Language Learners' Comparative Judgment on Text Difficulty

Introduction

1 Linguistic complexity (LC) needs to be carefully considered when selecting or adapting second language (L2) teaching materials, given its potential effect on text comprehensibility and processing (e.g., Bailey & Heritage, 2014) and the important role of comprehensible input in L2 acquisition (Gass, 2015). **2** Numerous readability formulas for text difficulty (TD) classification exist, mostly based on TD judgment data from first language (L1) speakers (see Benjamin, 2012, for a review). **3** Such formulas may be unreliable for selecting or adapting L2 teaching materials, as L1 and L2 speakers likely perceive TD differently given their vastly different language learning experiences (Nahatame, 2021).

4 Indeed, L1 and L2 processing have been found to be differentially affected by such variables as morphological regularity, syntactic patterns, semantic complexity (Goldschneider & DeKeyser, 2001), and frequency and association strength of phraseological units (Ellis et al., 2008). **5** According to the shallow structure hypothesis (Clahsen & Felser, 2006), during language processing L1 users utilize both syntactic computation (i.e., structural processing) and semantic and pragmatic information (i.e., shallow processing), whereas L2 users rely more on the latter and are thus likely less sensitive to syntactic cues than L1 users in reading (Marinis et al., 2005). **6** Furthermore, L2 reading is usually more cognitively effortful than L1 reading, especially for L2 readers with limited L2 knowledge and processing skills (e.g., Grabe, 2009). **7** Given the potential differences between L1 and L2 readers' perceptions of L2 TD, L2 researchers and teachers have been cautioned to avoid the comparative fallacy (i.e., expecting L1 and L2 users to process language in similar ways; Bley-Vroman, 1983) and to carefully consider L2 learners' needs in teaching material selection.

8 Although some readability models have been developed based on the cloze scores of learners of English as a foreign language (EFL; e.g., Crossley et al., 2008; Greenfield, 1999), it remains largely unknown how LC is associated with L2 learners' TD judgment. **9** Against this backdrop, our study examines the relationship of LC, measured using a comprehensive set of indices of lexical richness, syntactic complexity (SC), and discoursal complexity, on the one hand, and TD, captured by L2 learners' comparative judgment of comprehensibility and reading speed, on the other. **10** It is our hope that our findings can help to guide the selection and adaptation of L2 teaching material.

Earth	
Country	
Individual	

Identifying research gap(s)

Here you can give a brief overview of the current state of research on the topic, citing the most relevant literature. You can comprehensively analyze, summarize and comment on the research status of the research field (including the main academic viewpoints, previous research results and research level, the focus of the controversy, the existing problems and possible reasons, etc.), the new level, new dynamics, new technology and discovery, and the development prospects, etc., on the basis of the extensive reading and understanding of the literature in the research field involved in the topic after the topic has been determined. Comments that put forward their own insights and research ideas should be stated in a different style from the thesis. Or you can conduct a more in-depth survey of sources in the literature review section.

Depending on the specific field of study, your research may possess practical relevance, such as influencing policy decisions or aiding in effective management practices. However, it should be noted that its significance might predominantly lie in its contribution to the advancement of existing theories or the provision of new empirical data, appealing primarily to fellow researchers.

To make the relevance clear, explain how your paper helps solve a practical or theoretical problem, addresses a gap in the literature, builds on existing research and proposes a new understanding of the topic.

The main aim of this step is indicating how your work fits in. You need to explain your rationale for doing this research by relating it to existing work on the topic, contributing new insights, stating its broader social or practical relevance and pointing out why it matters.

Task 3.5 Read the following Introduction section and answer the questions.

The language of suffering: Media discourse and public attitudes towards the MH17 air tragedy in Malaysia and the UK

Introduction

In 2014, there were 20,218 documented civilian deaths in the Iraq war (Statista, 2018a), 7,823 people were killed by natural disasters (Guha-Sapir et al., 2015) and 1,328 people died in the air crashes, including the dramatic loss of life in the Malaysian Airlines flight MH17. These numbers reflect extreme human suffering around the globe. Acknowledging the rise of global suffering, scholars from different disciplines have, in recent years, investigated the relationship between suffering and media (Joye, 2012). This increasing academic attention has resulted in a wide range of research foci, such as the representation of distant suffering (e.g. Boltanski, 1999; Chouliaraki, 2011; Joye, 2009), media witnessing (e.g. Höijer, 2004; Kyriakidou, 2015), recovery discourse (e.g. Bonanno et al., 2010; Cox et al., 2008) and audience reaction towards mediated suffering (e.g. Huiberts and Joye, 2017; Seu, 2015).

While recognizing the diversity of research and richness of the ongoing academic research, we currently know little about the representation of air disasters within the media. However, according to a recent report published by Air Transport Action Group, air transport is one of the world's most important industries and offers important social benefits by providing the only transportation means in remote areas. In 2017, commercial airlines carried nearly 4 billion passengers, equivalent to half the world's population (Statista, 2018b). Thus, the investigation of aircraft accidents is important since it helps to understand human suffering in relation to the world's most important transportation industries. Specifically, this study aims to determine the ways in which the MH17 tragedy is linguistically defined and constructed in terms of keywords within both the media and the public context.

1. What's the research gap of this Introduction section?

2. What new insights will it contribute?

3. Does it have broader social or practical relevance?

4. In short, why does it matter?

Stating your aims and objectives

Many conventional forms of paper writing start with the discovery of a real problem as the starting point of the research. In the formal specification of dissertation writing, we generally follow the logical trilogy of writing, i.e., "formulate the problem–analyze the problem–solve the problem." When posing a problem, many dissertation writers analyze it in terms of a realistic research question. Generally, they present the real problems by describing the current situation, then analyze these problems and finally propose solutions. When it comes to effectively presenting genuine problems in research articles, numerous authors face constraints imposed by their disciplines or limited scientific research experience. Consequently, when composing such articles, they may inadvertently depict the current state or existing issues of the subject of study without substantial evidence, thereby introducing a notable degree of subjectivity into their work.

The research question needs to meet four criteria. On the top, research issues should be focused. First, the core of your research should always be centered on the research question to ensure that your research efforts are focused. If you have multiple questions, each of these research questions should be clearly linked to your core objective. Second, reliable sources should be used for your responses. Your questions must be answered using quantitative or qualitative data. Or your argument should be elaborated on by reading scholarly sources on the topic. Third, you should avoid subjective terms such as *good*, *bad*, *better*, *worse*, etc., as they do not give clear criteria for answering the question.

In addition, your research question should be feasible and specific. All terms in the research question should have a clear meaning. Vague language and overly broad ideas should be avoided. The research question does not require a conclusive solution or an action guide. The function of the research question is not indicative; it aims to promote understanding.

Besides, your research question should be complex and controversial. The research question cannot be answered merely with yes or no, because the yes/no category does not provide a broad enough scope for robust investigation and discussion. If readers can find the answer of the research question by doing a Baidu search, flipping through books and reviewing literature, then there is little point in posing this research question.

Last but not least, the research question should be relevant and original to the topic. Your research question should be developed based on initial readings on your topic. It should focus on addressing a problem or gap in existing knowledge in your field or course of study. It should contribute to a timely social discussion or academic debate that provides relevant knowledge for future researchers. The research question needs to be somewhat creative and should encompass inquiries that have not yet been definitively addressed.

Task 3.6 Read the following excerpts taken from two different research articles and answer the questions.

Text 1

History, modernity, and city branding in China: a multimodal critical discourse analysis of Xi'an's promotional videos on social media

The present study develops the multimodal study of city identity by looking at Chinese cities' digitalized branding practice through social media short videos. As noted by Powell et al. (2018, 579), short videos' dynamic visual flow can yield "a richer depiction of reality than static images." Despite the growing popularity of video-sharing social media platforms (e.g. TikTok, Snapchat Featured stories, Whatsapp's "moments" and Instagram's stories), to our knowledge, there has been no study that examines how the city branding discourse is shaped by the affordance of social media short video clips, let alone in the context of China.

Our study is premised on the notion of "symbolic economy" (Zukin et al. 1998), which identifies the nexus between urban policies and the material commodification of different aspects of culture. To map out the distinctive attributes that cities created on social media, we borrow the notion of "urban imaginary" proposed by Zukin et al. (1998) and Greenberg (2000), which refers to *a coherent, historically based ensemble of representations drawn from the architecture and street plans of the city, the art produced by its residents, and the images of and discourse on the city as seen, heard, or read in movies, on television, in magazines, and other forms of mass media* (Greenberg 2000, 228).

Adopting a critical multimodal discourse analysis method, we develop a semiotic framework to model Xi'an's digital image found in TikTok video clips as a set of evaluative attributes and to elucidate how they are constructed through verbal and visual resources. Our specific research questions include: (1) what are the distinctive urban imaginaries constructed in Xi'an's official TikTok videos, (2) how are the urban imaginaries realized through the use of linguistic and visual resources, and (3) what do the features of urban imaginaries reveal about urban policies in China and the influence of social media. In what follows, we will first introduce our theoretical basis of symbolic economy and city branding. We will then describe our data and analytical framework, which will be followed by an analysis of the urban imaginaries and their multimodal realization. Finally, the results will be discussed in relation to the current urban policies, affordances of social media, and implications for emerging second-tier cities' brands repositioning.

Text 2

Blessing or curse? Recontextualizing "996" in China's over work debate

Although "996" has been examined from different perspectives in different fields, such as legal studies, social psychology, human management and cultural studies (Hang, 2021; Wang, 2020), few studies have given a critical examination of the argumentative discourses at the core of the dispute and penetrate into the hegemonic struggles behind the dispute (Zhang, 2020; Zhen, 2021). This study views recontextualization as not only a representation of social events but also an appropriation of discourses across different social domains (Erjavec & Volčič, 2016; Fairclough, 2003), and aims to answer two research questions: (1) How are different discourses appropriated and manipulated by business tycoons and news media to (de)legitimate "996"? (2) What socio-political factors contribute to the hegemonic struggles behind the dispute?

1. What are the specific research questions raised by the authors?

2. How do the authors pose their research questions?

3. What are some common question words used to ask research questions?

LANGUAGE CONVENTION

Read the following sentences from the Introduction sections of the sample articles. Answer the following questions:

- **What function does Sentence 1 perform?**
- **What function does Sentence 2 perform?**
- **Why is the present tense used in Sentence 3?**
- **What function does Sentence 4 perform?**

1. However, despite the studies, we still lack an empirical understanding of how Chinese cities brand their identities, particularly using social media.

2. Addressing this need, the present study adopts a discourse analysis perspective and investigates the construction of city identities through systematic analysis of multi-semiotic resources in social media short videos.

3. The present study develops the multimodal study of city identity by looking at Chinese cities' digitalized branding practice through social media short videos.

4. Adopting a critical multimodal discourse analysis method, we develop a semiotic framework to model Xi'an's digital image found in TikTok video clips as a set of evaluative attributes and to elucidate how they are constructed through verbal and visual resources.

Giving definition

When writing about a topic, we need to clarify some terms, i.e., to explain clearly what theory and methodology you are going to use, especially in the research field of sociology and humanities. Thus, the section of Introduction may include a definition. Sometimes you need to give a definition yourself and sometimes you need to cite a proper definition from previous studies.

The first part may include notations and explanations of key words and technical terms, which are perhaps unknown to your readers, or there is a lack of agreement on. There are two types of definitions. One is called formal sentence definition; the other is known as extended definition.

An extended definition is longer than a short formal sentence definition and provides a fuller explanation of the concept. It usually begins with a general one-sentence definition and then provides additional details, such as types, history, examples, components, or applications. For example:

Term	Be	a/an/the	Category word	Distinguishing detail
Ego-network	*is*	*the*	*last level*	*focusing on the relationship between many structural properties of individual networks and individual attributes.*

It focuses on the relationship between many structural properties of individual networks and individual attributes which magnifies the individual core.

Task 3.7 Read the following definitions. Discuss with a partner how social network is defined in each of them.

1. **Social network** is an Internet based service that allows individuals to establish a public or semi-public personal homepage in a closed system to establish connections with other users. The core of social networks lies in the networking of relationships between people, manifested as a social network service platform built with various social networking software such as Blog App, WIKI App, Tag App, SNS App, RSS App, and other Web 2.0 core applications.

2. **Social network** is an Internet based service including hardware, software, services, and applications. Due to the fact that the four-character phrase is more in line with Chinese word formation habits, people tend to use social networks to refer to SNS (Social Network Service).

3. **Social networks** are gathering platforms of large information to acquire information and knowledge from which people can obtain useful information and knowledge to meet their learning needs and interests.

4. **Social networks** refer to Internet applications that provide communication and interaction services for online users in various forms, with certain social relationships or common interests as the link. This kind of social network mapping based on interpersonal relationship has formed a user centered and people-oriented Internet application on the Internet.

5. **Social network** refers to a way that enables people to establish and maintain social relationships online by means of information technology such as the Internet, which have also become an important way for people to access news, information, entertainment, and other aspects.

6. **Social network** is the product of the development of modern science and technology. It enables people to use the Internet and other tools to communicate with others online, share information, and establish social relations. The emergence of social networks has made it easier for people to connect with friends, family, and others, expanding the scope of interpersonal communication.

Task 3.8 Review the Sample Articles and try to give definitions of following items.

1. second language learners

2. discourse analysis

3. linguistic complexity

4. urban imaginary

Characteristic expressions

There are a number of expressions that are unique to the Introduction section. Try to familiarize yourself with them and pick some to use in your future writing. Expressions can be categorized to perform the following functions:

- Leading to a broad topic
- Claiming the centrality
- Reviewing related works
- Indicating the gap
- Introducing the present work

Scan the QR code for a list of the expressions.

Check your understanding

Task 3.9 Read the Introduction section in a research paper in your field and answer the following questions.

1. Is the Introduction section of the study clear and interesting? Do you feel like it deserves to be read in-depth?
2. Does the research direction of the paper follow naturally from the Introduction section and is this direction of enquiry fascinating?
3. Is there a review of relevant research trajectories and scholarly foundations that provides sufficient preparation for drawing out the ideas in this paper?
4. Is there a brief assessment and critique of existing research that summarizes the shortcomings?
5. Does it highlight the need and importance of the study from examining current research shortcomings?
6. Does it draw out the innovative and groundbreaking nature of this paper in the Introduction section?

Drafting Your Introduction Section

So far, you have learnt how to search for studies relevant to your own research interests, how to manage the retrieved literature, and how to write the introductory section of a research article. It is time to start drafting your own Introduction section to your chosen research topic. Please do the followings to complete this task.

Step 1: Find a topic of great research value that you are interested in.

Step 2: Choose 10–20 studies as references for your research proposal. List the references in the format required by your target journal.

Step 3: Compose an **OUTLINE** for the Introduction section. Indicate clearly the following:

- A general area related to your topic + known facts (*sources if necessary*)
- A subarea (*if there is one*) + known facts (*sources if necessary*)
- A specific area or your topic + previous studies (*references*)
- Centrality of the area / subarea / topic
- Research gap(s)
- Research question / aim / description
- Key points in your research methods

Step 4: Draft the Introduction section. Your Introduction section should be around 300–500 words long, containing essential information elements and citations introduced in this unit.

Unit 4

Describing Your Methods

Learning objectives

In this unit you will

- understand the general function of the Methods section;
- learn how to write the Methods section; and
- develop the linguistic strategies for writing an effective Methods section.

Self-evaluation

Scan the QR code, read the "Data and analytical method" section of Sample Article 6 (SA6) and answer the following questions.

- How do the authors start this section?
- How do the authors structure the elements of this section?
- How do the authors end this section?

In the journey of crafting impactful research articles in humanities and social sciences, drafting the Methods section is a critical step. This unit diverges from natural or experimental sciences by delving into a mix of methodologies like qualitative and quantitative research, including text analysis, case studies, ethnography, and more.

The core purpose of the Methods section is twofold: on the one hand, it provides a clear outline of the research process, detailing how data was gathered and analyzed; on the other hand, it ensures the study's integrity by making the research transparent and replicable. Such transparency plays a vital role in cultivating trust within the scholarly community and enables others to validate or expand upon the research findings.

Effective construction of this section involves a detailed yet concise narrative that covers research design, data sourcing, collection methods, and analysis strategies. It strikes a balance between detail for replication and readability for engagement.

Addressing ethical concerns, especially with human subjects, is of paramount importance. This unit offers guidance on clearly communicating these considerations, prioritizing participant rights and privacy.

The significance of a well-written Methods section cannot be overstated; it acts as the foundation of your research article. By detailing your methodology, this unit not only highlights your thorough approach to the research question but also lends credibility to your work. It encourages academic dialogue, enabling further research based on your methodological approach. This unit aims to guide you in creating a Methods section that reinforces the validity and significance of your research in the humanities and social sciences.

INFORMATION CONVENTION

Below is the simplified Methods section of Sample Article 6 (SA6). Read the provided text and identify the function of each paragraph.

History, modernity, and city branding in China: a multimodal critical discourse analysis of Xi'an's promotional videos on social media

Data and analytical method

1 The city under investigation is Xi'an, which was chosen for its successful rebranding of its city image by deploying rich symbolic resources and leveraging social media. As a traditional industrial city, Xi'an hit a bottleneck in the urban transformation process and has been seeking new urban imaginaries in response to intensified inter-urban competition in the new millennium. In 2018, Xi'an Tourism Bureau began to modernize its city image and established a full-scale cooperation plan with TikTok, a leading social media platform for creating and sharing short videos in China. The collaboration, which resulted in the effective exploitation of symbolic resources to transform its city image, made Xi'an "the most popular *wanghong* city" in China, surpassing tier-one cities like Beijing and Shanghai (Liu 2019). According to the *White paper on city image and short videos on social media*, up till December 2018, there were over 1.9 million TikTok videos related to the term "Xi'an" in hashtags, and the overall video views reached nearly 9 billion (Chinese Internet Data Centre 2018). The benefit of creating a *wanghong* image in the virtual world immediately manifested itself, with a 50% increase in tourist revenue in 2018 (Li and Jiang 2018). This study collected all the 294 short videos Xi'an Tourism Bureau posted on TikTok from May 2018 to May 2019, which was the most critical period during which Xi'an gained the reputation as the most popular *wanghong* city.

2 Our analysis of the city's urban imaginaries draws upon the method of Multimodal Critical Discourse Analysis (Machin and Mayr 2012). We consider city image as a set of evaluative attributes that are realized by different semiotic resources on the one hand and are shaped by the broader socio-cultural context on the other hand. We argue that cities' urban imaginaries are not "transparent" and do not "speak for themselves," rather, they are the result of complex signifying processes and should be interpreted based on the systematic analysis of the multimodal discourse (cf. Van Dijk 2007). Drawing upon Feng (2016), we propose a systematic framework to map out the multimodal realization of Xi'an's attributes, which is illustrated in Figure 1. The framework first distinguishes between attributes that are verbally articulated, either in the form of video subtitles or as utterances from characters, and attributes that are embedded in the visual depiction of sceneries and characters. Articulation can be further categorized into explicit and implicit ones. Explicit articulation refers to the use of attitudinal lexis (e.g. "fashionable," "stylish," and "splendid"), while implicit articulation resorts to facts or events that lead to a certain evaluation. For example, to represent Xi'an's international status, a video may explicitly label it as "an internationalized metropolis" or may articulate the evaluation implicitly by referring to the number of multinational companies in Xi'an.

semiotic 符号学的

3 Xi'an urban imaginaries are also constructed by visual depictions, which can be divided into depictions of sceneries and characters. Sceneries refer to Xi'an's scenic spots, including historical sites and cultural relics. As for the depiction of characters, we distinguish between the actional process and the analytical process, drawing upon Kress and Van Leeuwen (2006). The actional process refers to what a character is doing in the videos, such as dancing, singing, and playing musical instruments. The analytical process refers to parts that constitute a person as a whole, including a character's facial features, clothing, accessories, and other items that he/she "possesses."

actional 行动的

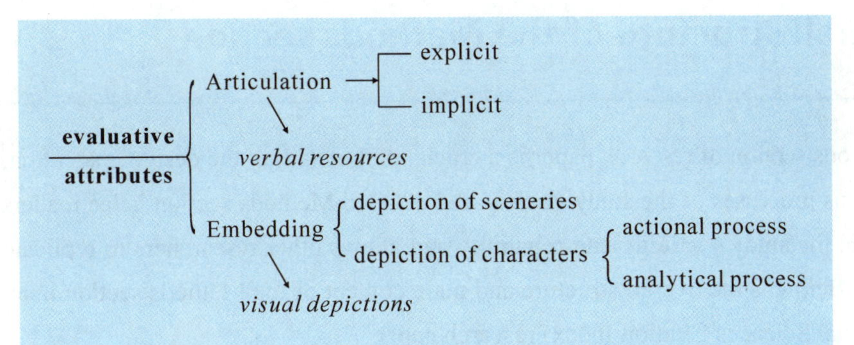

Figure 1. The multimodal construction of attributes.

4 The approach integrates the identification of attributes and the analysis of how they are discursively manifested, which makes the analysis more explicit and reliable (cf. Bateman and Schmidt 2012, 2). In the coding of the dataset, an attribute was counted each time when it was constructed through verbal text or visual image. When the same attribute is co-constructed by both verbal and visual resources in the same video, only one instance is recorded. There are also cases where different attributes are concurrently constructed in different semiotic modes. For example, in a video, the subtitle is about preserving Chinese cultural tradition (a traditional attribute), while a young girl dancing to some trendy music is represented visually (a modern attribute). As a result, the total number of attributes exceeds that of the videos collected. The two authors co-coded 20% of the videos independently and the results were compared. The agreement was above 80% and differences were resolved through discussion.

co-construct 共同构建

Paragraph	Function
1	
2	
3	
4	

The overall structure of the Methods section

The Methods section of research paper is a crucial part, detailing the design, execution, and analysis processes of the study. A clear and detailed Methods section helps readers understand the study's validity and reliability and allows other researchers to replicate the study. Below is the overall structure and main content of the Methods section in an SSCI (Social Sciences Citation Index) research paper.

1. Research design

Type of study: Specify whether the study is qualitative, quantitative, or mixed-methods.

Purpose of the study: Briefly state the problem being addressed or the phenomenon being explored.

Research framework: If applicable, describe the theoretical framework or conceptual model adopted for the study.

2. Sample or data sources

Sample selection: Explain the criteria for selecting the sample, the size of the sample, and how it was obtained.

Data sources: For studies based on existing data, it is crucial to provide comprehensive information about the source of the data and its relevance.

3. Data collection methods

Tools and techniques: Describe in detail the tools and techniques used for data collection, such as surveys, interviews, observations, etc.

Operational definitions: Provide clear operational definitions for key variables to ensure measurability.

Data collection process: Elaborate on the specific steps and procedures followed in collecting the data.

4. Data analysis methods

Analysis strategy: Clearly state how the data will be analyzed, based on the research question and type of data (e.g., statistical analysis, content analysis, thematic analysis, etc.).

Software utilized: Mention the software tools used for data analysis (e.g., SPSS, R, NVivo, etc.).

Statistical techniques: If conducting quantitative analysis, detail the statistical techniques and models used.

5. Reliability and validity of the study

Reliability: Discuss measures taken to ensure the consistency and repeatability in the data collection and analysis process.

Validity: Discuss how the findings effectively reflect the research question, including considerations of internal and external validity.

6. Ethical considerations

Participant consent: Informed consent was obtained from participants, who are directly involved individuals such as subjects, survey respondents, or interviewees. The consent process included explaining the study's purpose, methods, potential impacts, and confidentiality protections. Participants were informed of their rights, including the right to withdraw at any time. Consent was documented in writing or verbally with a witness, ensuring participants were fully aware and their rights protected.

Data confidentiality: Explain measures taken to protect participant privacy and the confidentiality of the data.

Ethical approval: If applicable, mention any ethical review approvals obtained for the research.

7. Limitations

Research limitations: Honestly discuss any potential limitations of the Methods section and their possible impact on the study findings.

A well-crafted Methods section not only provides enough detail to assess the quality of the research but also facilitate the study replication by fellow researchers, which is crucial for the accumulation and verification of scientific knowledge.

Task 4.1 Answer the questions below based on the above.

1. What are the key components of the research design section in an SSCI research paper's Methods section?
2. How should a researcher describe the sample or data sources in the Methods section?
3. What details should be included when describing data collection methods?
4. What constitutes the data analysis methods in the Methods section?
5. Why is it important to discuss the reliability and validity of the study in the Methods section?
6. What ethical considerations should be addressed in the Methods section?
7. How should researchers handle the limitations of their study in the Methods section?

Sample or Data Sources section

In crafting the Sample or Data Sources section of an academic research paper, a researcher lays the foundation for the study's transparency, reliability, and validity. This critical segment of the paper meticulously outlines the methodological choices made regarding sample selection and data gathering, ensuring that the research process is both clear and replicable for future scholars.

The journey begins with sample selection, where the researcher delineates the criteria and methodologies employed to select participants or data points. Whether it's through random sampling, convenience sampling, or stratified sampling, the rationale behind the choice and the process of determining the sample size are elucidated, often incorporating statistical methods or theoretical frameworks to justify these decisions.

Data sources are then scrutinized, with a thorough explanation of how existing data were sourced, including their creation, collection, and maintenance processes. The relevance of these data sources to the research question is underscored, highlighting their temporal and geographical pertinence to the study's aims.

Understanding the characteristics of the sample or data provides a window into the demographic details, composition, and specific contexts or populations represented. This section also addresses the representativeness of the sample or data set, discussing its ability to accurately reflect the target population and the implications for the study's findings.

Operational definitions play a pivotal role in ensuring clarity around the measures or variables used, facilitating a shared understanding and the potential for replication by other researchers. This clarity extends to ethical considerations, where the methods of obtaining informed consent from participants and safeguarding data confidentiality are laid out, reinforcing the study's ethical integrity.

Finally, a candid limitations statement offers a reflective examination of the sample or data sources' potential constraints, acknowledging how these limitations might influence the study's interpretative scope and its generalizability.

By meticulously detailing these elements, the researcher not only fortifies the study's methodological rigor but also contributes to the broader scholarly endeavor, paving the way for ongoing inquiry and the cumulative advancement of knowledge in the social sciences.

Task 4.2 Below is the Research Design section of Sample Article 5 (SA5). Based on the research design description provided, answer the following questions in brief. Each question is designed to test your understanding of the methodology employed in the study, particularly focusing on the rationale and structure of the mixed-methods approach.

Research Design

1 The study was conducted as a mixed-methods approach in which the data were collected by two types of questionnaires and 10 semi-structured interviews. **2** The rationale for using a combination of these approaches is that the quantitative data can provide readers with a comprehensive image of the research issue; qualitative data collection provides information that refines, extends, or explains the concept (Subedi, 2016). **3** More specifically, this study was designed in an explanatory sequential design, using qualitative data to interpret the results obtained from the questionnaires. **4** Therefore, combining these two methods ensures the validity and reliability of the results and enhances in-depth analysis. **5** Figure 1 displays the research procedures of this current study.

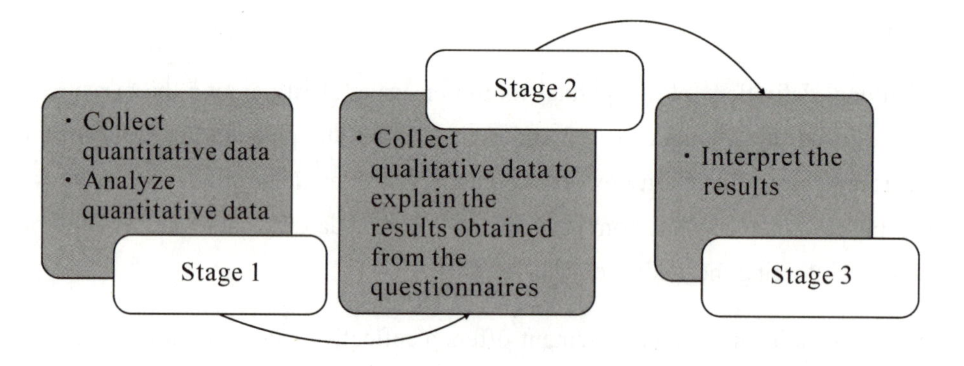

Figure 1. Research procedures

1. What two methods of data collection are employed in this mixed-methods study?

2. Why does the researcher choose to combine quantitative and qualitative methods in this study?

3. What specific design type is used to structure the research, and how does it organize the data collection process?

4. How do qualitative data contribute to the interpretation of the quantitative findings in this study?

5. What benefits does the combination of these two methods provide to the overall validity and reliability of the study's results?

Data collection methods

In the realm of scholarly research, particularly within the SSCI publications, the articulation of data collection methods stands as a cornerstone of methodological transparency. This section serves not just to illuminate the chosen apparatus of inquiry but also to underpin the study's integrity, facilitating a critical assessment of its reliability and validity. Equally, it paves the way for future endeavors in the field by offering a blueprint for replication or refinement.

At the heart of this narrative are the tools and techniques employed for data collection. Researchers meticulously detail the instruments—be it questionnaires, interview guides, or observation sheets—employed in the gathering of data, alongside a candid exposition of their development and validation processes, including pilot studies. The selection of data collection techniques, such as face-to-face interviews or online surveys, also needs to be justified. This justification provides valuable insight into the methodological choices that shape the research landscape.

Operational definitions play a pivotal role, offering clarity on key variables and ensuring the robustness of the measurement. This practice not only guarantees data consistency but also fortifies the study against methodological pitfalls.

The data collection process is laid out with precision, charting the logistical map of the research journey from the timing and locales of data collection to the strategies for participant engagement. Documentation practices are also disclosed, highlighting efforts to maintain data integrity.

Data quality control emerges as a thematic imperative, with researchers delineating the safeguards—like data verification and coding accuracy checks—put in place to vouchsafe the data's authenticity.

Ethical considerations are of paramount importance, with informed consent and privacy protections underscored, alongside any ethical review protocols the study may have undergone. This not only reflects adherence to ethical standards but also reinforces the moral fabric of the research.

Lastly, the narrative acknowledges the challenges encountered and surmounted, offering a transparent account of obstacles in data collection and the strategies devised to overcome them.

Through its detailed exposition, this section acts as a linchpin of scholarly inquiry, embodying the principles of transparency, rigor, and ethical conduct that define academic pursuit. It not only vouches for the study at hand but also contributes to the cumulative wisdom of the research community, fostering a culture of integrity and replicability.

Task 4.3 Below is the Text Selection part of Sample Article 1 (SA1). Based on the detailed description provided about the selection of texts within a specific research context, please answer the following questions to demonstrate your comprehension of the methodologies and criteria used in the study.

Text Selection

This study sampled 74 narrative and 106 expository texts from Books 2–4 of the four-book New Concept English coursebook series, which have been widely used by non-English-major EFL learners in China. These books were compiled for beginning (Book 2), intermediate (Book 3), and advanced (Book 3) EFL learners, respectively, with a total of 204 texts (word range: 95–720). Specifically, we sampled 70 texts from Book 2 (61 narrative and 9 expository), 60 from Book 3 (13 narrative and 47 expository), and 50 from Book 4 (50 expository) to represent a range of TD. As these texts varied greatly in length (word range: 148–720), we broke down each of the 110 texts with over 200 words into excerpts at natural points and used the beginning excerpts of those texts (word range: 153–165) along with the 70 unabridged texts (word range: 148–177) for a pilot test. We invited seven raters who met our rater recruitment criteria (see the next section) but did not participate in the actual study to rate the TD of these texts, with each rater making about 50 pairwise comparisons. The raters indicated that they tended to consider longer texts to be harder to comprehend and/or process than shorter texts. Based on their feedback, we set the word length of all texts uniformly at 150 words to further control for text length as an intervening variable. To this end, we invited two L1 English speakers to revise the final one or two sentences of the 110 text excerpts and the 70 unabridged texts.

The texts covered topics on arts, politics, education, and astronomy. We invited 10 learners who met our rater recruitment requirements but did not participate in the actual study to rate their familiarity with the topics of the 180 texts on a five-point scale (with 1 indicating very unfamiliar and 5 indicating very familiar). Their ratings ranged from 3 to 5 ($M = 3.881$; $SD = 0.246$), indicating that the topics were familiar to our raters. The mean familiarity ratings were not significantly correlated with L2 raters' CJ scores on comprehensibility ($r = .066$, $p = .384$) or reading speed ($r = -.048$, $p = .524$; see the Results section for CJ scores). In addition, the first researcher and three of the 10 pilot topic-familiarity raters read all abridged texts and unanimously agreed that all texts had coherent meanings.

1. From which books in the New Concept English coursebook series were the texts sampled for the study?
   A. Books 1–3.
   B. Books 2–4.
   C. Books 1–4.
   D. Books 3–4.

2. What was the primary basis for breaking down the longer texts into excerpts?
   A. The genre of the text.
   B. The topic of the text.
   C. The word range.
   D. The complexity of vocabulary.

3. How did the raters indicate that text length influenced text difficulty?
   A. Shorter texts were easier to comprehend.
   B. Longer texts were easier to process.
   C. Longer texts were harder to comprehend and/or process.
   D. Text length had no impact on comprehension.

4. What was the uniform word length set for all texts after feedback from the pilot test?
   A. 100 words.
   B. 150 words.
   C. 200 words.
   D. 250 words.

5. How familiar were the raters with the topics of the 180 texts based on their ratings?

A. 1 to 3.

B. 1 to 5.

C. 3 to 5.

D. 2 to 4.

Data analysis methods

In the crafting of an academic research paper, the elucidation of data analysis methods within the Methods section is a pivotal endeavor. This narrative not only demystifies the journey from raw data to polished insights, but also fortifies the study's standing in terms of its reliability, validity, and the potential for future replication.

The tale begins with data preprocessing, a meticulous phase where data are cleaned, normalized, and perhaps transformed, being prepared for the analytical spotlight. This phase sets the stage for integrity and precision, ensuring that the subsequent analysis rests on a solid foundation.

As the plot unfolds, the chosen analytical strategies take center stage. Whether the narrative is woven through quantitative threads—employing statistical models and inferential techniques—or through qualitative patterns, using thematic or discourse analysis—the methods employed are detailed with precision. The selection of specific software tools, such as SPSS for statistical analysis or NVivo for qualitative data coding, should also be justified.

The development of statistical models or the application of qualitative frameworks is then narrated, highlighting the methodological rigor and theoretical underpinnings that guide the analysis. This is where the study's hypotheses are tested against the empirical evidence, a critical juncture in the research journey.

It should be noted that a research narrative is incomplete without addressing the challenges and ethical considerations that accompany data analysis. From ensuring data security and participant privacy to navigating analytical hurdles, the story delves into the measures taken to uphold the ethical standards and methodological integrity of the study.

In conclusion, the part of data analysis methods is more than just a methodological exposition; it is a testament to the study's credibility, offering a transparent window into the analytical soul of the research. By detailing the analytical journey with clarity and precision, it not only paves the way for scholarly assessment, but also invites future scholars to tread the same path, thereby enriching the tapestry of scientific inquiry.

Task 4.4 Below is the Data Analysis section of Sample Article 1 (SA1). Based on the detailed description provided about the statistical models and analytical procedures employed within the specific research context, please answer the following questions to demonstrate your understanding of the methodologies and criteria used in the study. The questions will assess your comprehension of the key steps, models, and validation techniques applied in the analysis.

Data Analysis

The NMM system automatically generated standardized estimates of the TD levels of the texts (henceforth CJ scores on comprehensibility and reading speed) from the L2 raters' binary CJ decisions for comprehensibility and reading speed by means of the Bradley–Terry–Luce model (Bradley & Terry, 1952; Luce, 1959), a type of logistic model for predicting the likelihood of one item in a paired comparison being chosen over the other that has been commonly used in CJ studies (e.g., Crossley et al., 2017; Crossley, Skalicky, & Dascalu, 2019; Paquot et al., 2022). To test the relationship between the LC indices and L2 raters' CJ on TD, we developed regression models with the LC indices as independent variables to predict the CJ scores on comprehensibility and reading speed, respectively. To this end, the LC indices that violated a normal distribution were discarded first. Second, a Pearson correlation analysis was carried out to determine whether the remaining indices were correlated significantly ($p < .05$) and meaningfully (i.e., $|r| \geq .10$, representing at least a small effect size; see Cohen, 1988) with the CJ scores on comprehensibility and reading speed. The indices without a significant and meaningful correlation were removed from further consideration. Third, multicollinearity among the LC indices was checked. For each pair of indices with a correlation coefficient of .8 or larger, the index with a weaker correlation with the CJ scores was discarded.

Fourth, stepwise linear regression models with the retained LC indices as predictor variables and the CJ scores on comprehensibility or reading speed as a dependent variable were built using the Akaike information criterion method. Fifth, a follow-up 10-fold cross-validation was performed using the LC indices in the obtained regression models to test whether they were consistent across the data set. Sixth, as it is not always reliable to use standardized beta estimates to determine the relative importance of predictor variables in regression models (see Mizumoto, 2023, for review), we conducted relative weight analyses for our final models via Mizumoto's (2023) R-based web application (http://langtest.jp/shiny/relimp) to disassemble the variance explained by the predictors and assess their relative importance (Tonidandel & LeBreton, 2011). Finally, we employed Lenhard and Lenhard's (2014) calculator to perform Fisher *r*-to-*z* transformation to test the differences between the correlations obtained from our models and two readability formulas. The data and the R scripts used for data analysis have been made publicly available through the Open Science Framework.

1. Which model was used to generate the standardized estimates of TD levels in the study?
   A. Ordinary Least Squares (OLS) Regression.
   B. Bradley–Terry–Luce model.
   C. Poisson Regression Model.
   D. Hierarchical Linear Model (HLM).

2. What was the first step taken to test the relationship between LC indices and CJ scores on TD?
   A. Performing a Pearson correlation analysis.
   B. Checking multicollinearity among LC indices.
   C. Discarding LC indices that violated normal distribution.
   D. Conducting stepwise linear regression models.

3. How was multicollinearity among LC indices addressed in the study?
   A. By using factor analysis to combine variables.
   B. By discarding the index with a weaker correlation when two indices had a correlation coefficient of .8 or larger.

C. By retaining all indices for further analysis.

D. By performing a hierarchical regression model.

4. What method was used to build the regression models in the study?

A. Bayesian information criterion.

B. Cross-validation method.

C. Akaike information criterion.

D. Principal component analysis.

5. Why did the researchers perform a 10-fold cross-validation on the regression models?

A. To refine the predictor variables.

B. To test the consistency of the models across the dataset.

C. To increase the sample size for analysis.

D. To eliminate multicollinearity issues.

6. How did the researchers assess the relative importance of predictor variables in the regression models?

A. By using standardized beta estimates.

B. By conducting relative weight analysis using Mizumoto's R-based web application.

C. By using variance inflation factors (VIF).

D. By performing stepwise logistic regression.

LANGUAGE CONVENTION

Read the following sentences from the Methods sections of Sample Article 2 (SA2). Answer the questions below:

- **Why is past tense used in Sentence 1?**
- **What is the purpose of using passive voice in Sentence 2?**
- **Why is active voice used in Sentence 3?**
- **What is the purpose of using technical vocabulary in Sentence 4?**
- **Why is Sentence 5 complex?**
- **Why does Sentence 6 mention a lot of details?**
- **How does Sentence 7 reflect the study's fairness and methodological rigor?**

1. I designed my search to minimize the likelihood that it would fail to find a supplementation-relevant article from the Common Core era, but the wide net I cast ensured that my search would yield many irrelevant articles.

2. The ubiquity of virtual resource pools can be linked to Common Core–driven standardization...

3. After this initial screening, I performed a "snowballing" search of the reference sections of articles retained after screening (Greenhalgh & Peacock, 2005), this time relaxing the restriction that articles be published in 2015 or later.

4. I began with searches of the reference databases Education Resources Information Center (ERIC) and PsycInfo, restricted to 2015 and later.

5. For the purposes of the first two questions, then, I systematically reviewed literature from this most recent era of curriculum supplementation, from 2015 to December of 2020, when I ran my final systematic literature search.

6. After screening and unduplicating results returned in both the ERIC and PsycInfo searches, there were 64 articles retained in my collection of Common Core era, supplementation-relevant work.

7. The first criterion is a necessity because English is the only language in which I am fluent.

Using appropriate tenses

In writing the Methods section of an academic paper, it is crucial to use the correct tense. Generally, this section should predominantly use the past tense, as it describes the research process and methods that have been completed. Here are specific recommendations for tense usage:

- **Past tense:** When describing the experimental steps you took, observations made, data collection methods, and the research design implemented, you should use the past tense. This is because these activities have been completed at the time of writing the paper. For example: "We conducted a survey" or "We analyzed the data using...".

- **Past perfect tense:** In some cases, if you need to emphasize that an action was completed before another action, the past perfect tense can be used. This is particularly useful when describing a series of steps that occurred in sequence. For example: "By the time we started the analysis, we had collected all necessary data."
- **Present tense:** Although most of the method description should be in the past tense, the present tense is expected to be used when discussing facts that are universally accepted as true or when describing the general validity of the study findings. Additionally, the present tense should be used when citing literature or explaining methods that are still valid at the present time. For example: "The XYZ technique allows for precise measurement of...".

In summary, the Methods section should primarily employ the past tense to detail the specific operations and processes of the research. This reflects the fact that the study has been completed and helps readers understand exactly how the research was conducted. However, depending on the context, the appropriate use of the past perfect tense or present tense is also necessary. Correct tense usage not only improves the clarity of the article but is also part of academic writing standards.

Task 4.5 Read the following sentences. Discuss with a partner the reason why a particular tense is adopted.

1. The city under investigation is Xi'an, which was chosen for its successful rebranding of its city image by deploying rich symbolic resources and leveraging social media.

2. Xi'an hit a bottleneck in the urban transformation process and has been seeking new urban imaginaries in response to intensified inter-urban competition in the new millennium.

3. In 2018, Xi'an Tourism Bureau began to modernize its city image and established a full-scale cooperation plan with TikTok.

4. The collaboration, which resulted in effective exploitations of symbolic resources, made Xi'an "the most popular *wanghong* city" in China.

5. Our analysis of the city's urban imaginaries draws upon the method of Multimodal Critical Discourse Analysis (Machin and Mayr, 2012).

6. We propose a systematic framework to map out the multimodal realization of Xi'an's attributes.

Using the appropriate voice

When writing the Methods section of an academic paper, the voice used primarily depends on how you want to present your research process. Typically, this section aims to detail how the research was designed, conducted, and analyzed so that other researchers can understand and replicate your study. The voice used in the Methods section can be either active or passive, each with its advantages and appropriate contexts:

- **Active voice:** Using the active voice can make the narrative more direct and dynamic. It clearly identifies the doer of the action (usually the researchers). In certain cases, employing the active voice can make the text easier to read and understand, especially when it's important to highlight the contribution of the researchers. For example: "We surveyed 500 respondents to collect data." The active voice in this sentence makes the actors and their actions more apparent.
- **Passive voice:** In many scientific writing and academic writing traditions, the passive voice is widely used, particularly in the Methods section. Using the passive voice can make the description more objective by focusing on the action itself rather than the doer of the action. This helps to maintain the objectivity and standardization of the text. For example, "500 respondents were surveyed to collect data." This expression focuses attention on the research process, not the researchers conducting the process.

The choice between which voice to use should be based on the following factors:

- **Guidelines and conventions of the journal:** Different academic journals and fields may have different preferences for voice. Checking the author guidelines of your target journal is important before preparing your manuscript.

- **Clarity and accuracy:** Choose the voice that conveys your research methods most clearly and accurately.
- **Traditions of the research and expectations of the readers:** Consider the traditions in your field of study and what your target readers might expect in terms of voice.

Overall, while the passive voice is more common in the Methods section, the active voice can also be used depending on the specific needs of the article and the requirements of the journal. Most importantly, regardless of the chosen voice, ensure that the Methods section is clear, accurate, and easy for readers to understand and replicate.

Task 4.6 Read the following sentences. Discuss with a partner the reason why a particular voice is chosen.

1. The gross domestic product (GDP), trade openness, applied patents (Patents, number of total patents), and renewable energy consumption were taken from the World Bank.
2. Energy consumption and energy intensity were taken from the Enerdata Yearbook.
3. The ecological footprint and load capacity factor were taken from the Global Footprint Network.
4. The convergence methodology of Phillips and Sul is also applied.
5. The Westerlund cointegration test can be applied.
6. In this research, PCSE and FGLS will be used to verify the robustness of the results.
7. The club convergence test is applied for the analysis of conditional and absolute convergence.
8. The causality test is applied to analyze the predictive capacity of one variable over another.

Understanding nominalization

The phenomenon of nominalization in the Methods section of academic papers is a common linguistic style in academic writing, especially when describing research methodologies and processes. Nominalization involves transforming verbs, adjectives, or other parts of speech into nouns, which serves several key purposes and advantages:

- **Enhancing professionalism and precision:** Through nominalization, authors can discuss specific research methods, theoretical concepts, and practices in a more professional and precise manner. For example, by transforming the term "measure" into "measurement," the emphasis shifts towards the process or tool utilized for measuring, rather than solely focusing on the act of measurement. This linguistic transformation underscores the significance of the methodology employed in the research, highlighting its central role in the study.
- **Emphasizing concepts and processes:** Nominalization helps highlight key concepts, theories, and processes in the research, rather than the actions themselves. This style makes the text more objective and depersonalized, contributing to the emphasis on the universality and replicability of the research. For instance, using "analysis" instead of "analyze" underlines the importance of the analytical process.
- **Increasing text compactness:** The utilization of nominalization in academic articles enables authors to succinctly convey intricate concepts and information, thus facilitating brevity and enhancing readability. This linguistic technique plays a vital role in condensing complex ideas without compromising the clarity and comprehensibility of scholarly writing. By focusing on descriptions, more information can be conveyed in limited space.
- **Facilitating academic communication:** Within the academic community, nominalization establishes a common professional language, enabling research methods and results to be understood and discussed across disciplines and cultures. This standardized way of expression promotes the dissemination and exchange of academic knowledge.

- **Addressing research complexity:** Academic papers often involve complex theories and methodologies, and nominalization enables authors to manage this complexity more effectively. By clearly referencing key concepts and methods, nominalization supports the arguments and conclusions of the paper.

In summary, the widespread use of nominalization in the Methods section of academic papers reflects the need for precision, objectivity, and efficiency in academic writing. By adopting noun forms, research methods and concepts are clearly presented, aiding readers in understanding the design, implementation, and evaluation of the study.

Task 4.7 Use nominalization to transcribe the following sentences in the context of describing research methodologies and processes.

1. We analyzed the data using statistical methods.

2. The researchers collected samples from different locations.

3. They measured the temperature daily.

4. The study explores the impact of social media on self-esteem.

5. We surveyed 200 participants for the study.

6. The researchers observed the behavior of participants throughout the experiment.

7. The team recorded responses using a standardized questionnaire.

8. They calculated the average scores for each group.

9. The researchers interviewed the subjects after the experiment.

10. They compared the results with previous studies.

Characteristic expressions

Some expressions are characteristic of the Methods section of an academic paper. Try to get familiar with them and pick some to use in your future writing. The expressions can be categorized as follows:

- Describing the research design and approach
- Sampling and participants
- Data collection methods
- Instrumentation and materials
- Data analysis techniques
- Ethical considerations
- Limitations and scope

Scan the QR code for a list of the expressions.

Task 4.8 Please select the most appropriate option for each question.

1. Which of the following sampling techniques ensures a representative distribution of the population?

   A. Stratified sampling.

   B. Random sampling.

   C. Convenience sampling.

   D. Purposive sampling.

2. What is the primary purpose of employing a cross-sectional design in research?

   A. To capture specific aspects at a single point in time.

   B. To explore changes and developments over time.

   C. To examine the effects of an intervention.

   D. To provide a detailed examination of a single case.

3. Which method is specifically chosen for its effectiveness in obtaining detailed and in-depth information?
   A. Structured questionnaire.
   B. Semi-structured interviews.
   C. Observational checklist.
   D. Archival analysis.

4. The rationale behind choosing a specific design is influenced by its suitability for:
   A. Exploring.
   B. Analyzing.
   C. Examining.
   D. All of the above.

5. In ensuring data quality and reliability, what is a crucial step for survey responses?
   A. Conducting a power analysis.
   B. Pilot-testing the questionnaire.
   C. Using a systematic sampling method.
   D. Employing multiple observers.

6. Ethical considerations in research primarily include:
   A. Increasing sample size.
   B. Using advanced statistical software.
   C. Obtaining informed consent and ensuring confidentiality.
   D. Publishing results quickly.

Check your understanding

Task 4.9 Analyze the Methods section in the Sample Articles and finish the tasks below.

1. Choose one of the provided Sample Articles and identify its research design (e.g., quantitative, qualitative, mixed methods). Summarize the reasons for choosing this design and discuss how it supports the resolution of the research question or objective.

2. Locate the description of the sampling technique (such as random, convenience, stratified sampling) in a selected sample article. Evaluate the rationale behind the author's choice of this sampling method and discuss its potential impact on the representativeness and validity of the research findings.

3. Analyze the data collection methods described in a sample article. Identify the tools used (e.g., questionnaires, interview guides, observation checklists) and the collection procedures. Assess the effectiveness of these methods in gathering relevant, reliable data.

4. Find and summarize the data analysis methods (such as descriptive statistical analysis, thematic analysis, regression analysis) in a sample article. Explore how the author chose specific analysis techniques based on the research design and type of data, and evaluate the appropriateness of these choices.

5. Identify and discuss the ethical considerations mentioned in a sample article, including but not limited to the acquisition of informed consent, protection of participant privacy, and secure handling of data. Analyze the adequacy of these ethical measures and their potential impact on research participants and outcomes.

6. Extract the research limitations section from your chosen sample article. Summarize the limitations identified by the author and reflect on how these limitations affect the interpretability and generalizability of the research. Based on these limitations, propose potential directions for future research.

Unit task

Drafting Your Methods Section

After mastering the art of literature search, literature management, and drafting the Introduction section of a research article, it's time to draft the Methods section of your research proposal. This section is crucial as it outlines how you plan to conduct your study, ensuring that your research is replicable and transparent. Follow these steps to complete this task.

Step 1: Select 10–20 methodological studies as references.
Identify and choose 10–20 methodological studies that are relevant to your research proposal.

These studies should serve as a foundation for the design, sample selection, data collection, and analysis techniques you intend to employ in your study. List these references in the format required by your target journal, ensuring to include both seminal works and recent studies that can justify your methodological choices.

Step 2: Compose an outline for the Methods section.

Craft a detailed outline for your Methods section, covering the essential components that will guide your research execution:

- **Research design:** Describe the overarching approach (qualitative, quantitative, mixed methods) and the specific type of study (experimental, survey, case study, etc.). Justify your choice of design in relation to your research question.
- **Sample or data sources:** Detail your target population and the sampling method you will use (e.g., convenience, stratified random sampling). Define inclusion and exclusion criteria, and justify the size of your sample.
- **Data collection methods:** List the instruments (surveys, interview guides) and procedures (online, in-person) for data collection. Mention any pilot testing or pre-validation of tools.
- **Data analysis methods:** Outline how you will analyze the data (statistical tests, thematic analysis) and any software that will be used. Mention measures for ensuring data quality and integrity.
- **Reliability and validity:** Discuss how you will ensure the reliability (e.g., test-retest, internal consistency) and validity (content, construct, criterion) of your study.
- **Ethical considerations:** Outline how you will address ethical issues, including informed consent, confidentiality, and potential risks to participants.
- **Limitations:** Acknowledge any potential limitations of your methodology and their implications for your findings.

Step 3: Draft the Methods section.

Using your outline as a guide, draft the Methods section of your research proposal. This draft should be concise yet comprehensive, ranging from 300 to 500 words. It must clearly communicate how you intend to carry out your study, ensuring that your methods are robust, ethical, and aligned with your research objectives. Incorporate references from your selected methodological studies to substantiate your choices and highlight the rigour of your approach. Ensure that your draft

- is logically organized, following the structure outlined in step 2;
- clearly justifies each methodological choice in relation to your research question and objectives;
- includes citations from your list of methodological studies to support your decisions;
- addresses potential ethical issues and how they will be mitigated; and
- is written clearly and concisely, adhering to the style guide of your target journal.

Unit 5

Presenting Your Results

Learning objectives

In this unit you will

- understand the general function of the Results section;
- learn how to write the Results section; and
- understand the linguistic strategies for writing an effective Results section.

Self-evaluation

Scan the QR code, read the Results section of Sample Article 1 (SA1) and answer the following questions.

- What kind of information should be covered in the Results section?
- How do the authors restate the method-related details in the Results section?
- How do the authors report the results?
- How do the authors interpret and comment on the results?
- How do the authors use visual aids to present the results?

A research paper's main goal is to close a gap in the literature on a specific subject. In light of this, the paper's Results section—which outlines the main conclusions of the investigation—is sometimes regarded as its central body. Reviewers, peers, readers, and any news outlets covering your research focus the most on this part of your work. Partially due to this, one of the most crucial aspects of preparing your research paper is writing a Results section that is clear, simple, and logical.

Firstly, you should give a summary of the experiments in the Results section and, if needed, use tables and graphs to arrange the data you collected logically. Put differently, the study's conclusions must be covered in depth in the Results section. However, making linkages between the various findings and comparing them to earlier discoveries in the literature is not the purpose of this section; they are the domain of the Discussion section.

The format of the Results and Discussion sections in papers might vary depending on the discipline, journal, and type of research. The Results and Discussion sections can be either combined together or prepared separately.

Secondly, your findings should be presented objectively and without bias in the Results section. The Results section must be strictly factual, in contrast to the Discussion section, which may discuss hypothetical topics. For instance, you might have discovered a strong correlation in your research between two variables that had never been reported before. It makes sense to clarify this in the Results section. Speculation regarding the causes of this correlation, however, belongs in your paper's Discussion section.

Thirdly, you'll probably be working with statistical analysis results if you did quantitative research. The outcomes of any statistical tests you performed to compare groups or evaluate the relationships between variables should be included in your Results section. It should also indicate if each hypothesis was validated or not.

Graphs, charts, and tables are examples of visual elements that are frequently useful in quantitative research; nevertheless, they should only be included if they are directly related to your findings. To make it easier for your reader to grasp what is being presented, give these elements titles and labels that are clear and informative. Consider including a figure and table list if you would like to include any more tangential graphic components.

Visual components that summarize and show the findings include figures, charts, maps, tables, and so forth. These components need to be numbered sequentially and cited inside the text. The legend should have sufficient information to comprehend the non-textual part if figures and tables are to be able to stand alone without the text.

INFORMATION CONVENTION

Below is the excerpt of Sample Article 6 (SA6) about city image construction. Read it and fill in the table with the numbers of the sentences that include the given elements.

1 Based on the detailed analysis of the 294 videos, we identified a clear set of attributes that Xi'an uses to construct its urban imaginaries. **2** The imaginary includes the image of a modern metropolis and the image of a historical city, as shown in Table 1.

Table 1. The distribution of modern and historical attributes

Modern metropolis (197)	Historical city (136)
Stylishness (43)	Recreating the Tang Dynasty (89)
Youthfulness (39)	Revitalizing folk art (47)
Fantasy city (37)	
Popularity (33)	
Internationalness (45)	

3 The modern city image is created through using *wanghong* rhetorical strategies, which are propelled by the global attention economy (Marwick 2013). **4** The city is personified as a popular microcelebrity who uses TikTok to create a set of popular imaginaries, including stylishness, youthfulness, fantasy city, popularity, and internationalness. **5** These imaginaries allow Xi'an to abandon its original boring functional attractions and to create a landscape of mass entertainment, tourism and investment. **6** The historical city image is constructed by representing Xi'an as an ancient capital, which includes exploiting the cultural symbols of the Great Tang Dynasty and local folk art. **7** The Great Tang Dynasty is recreated through the appreciation of the past history, culture and landscape of the past, and local folk art (e.g. shadow-puppets, paper-cutting, worship parade). **8** By creating a series of nostalgia-oriented, traditional cultural symbols, the idealized urban imaginary of "Great Tang" is combined with utopian visions of Xi'an's future growth. **9** In what follows, we will provide a detailed analysis of how these attributes are discursively constructed in the official TikTok videos.

Elements	Sentence No.
A very short introductory context that repeats the research question	
Report on data collection, recruitment, and/or participants	
A systematic description of the main findings in a logical order (generally following the order of the Methods section)	
Visual elements, such as figures, charts, maps, tables, etc. that summarize and illustrate the findings	

The overall structure of the Results section

Depending on the type of study done, the format of the research Results section may change, but generally speaking, it should contain the following elements.

Introduction: A generalization of the study's objectives, goals, and research questions should be included in the introduction. It should also provide a brief explanation of the study's methodology.

Data presentation: The data gathered for the research is shown in this part. Tables, graphs, and other visual aids may be included to help readers comprehend the material.

Headings and subheadings should be used to help the reader navigate the thorough logical and coherent organisation of the facts presented.

Data analysis: The data from the previous section are examined and explained in this section. The findings of the statistical tests should be presented in an understandable way, and the tests themselves should have been interpreted in a clear manner.

Results discussion: In this section, the study's findings should be interpreted and any unexpected discoveries should be discussed. The research questions of the study and explanation of how the findings advance the field of study should also be covered in the discussion.

Limitations: Any study restrictions, including sample size, data collection techniques, or other elements that might have affected the outcomes, should be mentioned in this area.

Conclusions: Summarizing the major discoveries of the research and a final analysis of the outcomes should be included in the section. The research objectives of the study and explanation of how the findings advance the field of study should be included.

Recommendations: Drawing from the study's findings, recommendations for further research should be provided in this section. It might also point to useful practical implicature of the study's findings in actual contexts.

Note: Not all the information elements may appear in a paper. There are disciplinary variations. The elements may not necessarily follow the given order. Some may appear in cycles. Some may be integrated and not clear-cut from each other.

Task 5.1 Read the Results section of SA6 again and answer the questions below.

1. What is the research objective of the study?

2. What are the key findings of the sample article?

3. The excerpt is the first subsection of the Results section. Can you predict the next subheading?

Restating method-related details

An introduction and a restatement of the research question should be included, even though the Results part of a research paper primarily reports the findings. This connects with the paper's preceding sections and ensures a coherent flow of information.

You should first describe the flow of participants each step along the way in your research and whether any data were eliminated from the final analysis before starting with your research findings.

Recruitment period and participant flow

Any attrition, or the drop in participants at each subsequent stage of a study, must be reported. The reasoning behind this is that any unequal distribution of participants among groups can occasionally jeopardize internal validity and complicate group comparisons. Remember to list every cause of attrition as well. For instance:

Out of the 298 individuals who finished the preliminary screening questionnaire, 78 (26.1%) were deemed unsuitable for inclusion in the study due to their non-compliance with caffeine consumption (11%) or prescription drug use (15.1%).

In order to receive the credit for the research, the remaining 220 participants were asked to finish the online study survey. Nevertheless, twelve more individuals were unable to finish it, consequently, there was a final count of 208 participants.

The ideal way to report the number of participants in each group per stage and the causes for attrition is using a flow chart, especially if your study has several groups (experimental and control groups) and stages (pre-test, intervention, and post-test).

Report the dates at which you recruited participants of the study or follow up with new sessions.

Missing data

The completeness of your datasets is another vital concern. Reporting of both the quantity and justifications for missing or excluded data is expected.

Unexpected events, participant ineligibility, equipment failures, inappropriate storage, and other factors can render data unusable. Indicate the reason the data were useless for each instance. Because they are outliers, some data points might be eliminated from the final analysis; however, you must be able to explain your decision-making process. For example:

Thirteen individuals' data were eliminated due to their inaccurate responses to the attentiveness check question. Due to equipment malfunctions, data for two more individuals were lost.

Please report any strategies you used to overcome or make up for missing data as well.

Adverse incidents

Report any adverse effects or occurrences with major repercussions that occurred for clinical research.

Task 5.2 **Read the following part of the Results section about corporate social responsibility and identify the participant flow and recruitment period, the missing data, and the adverse events. Complete the table under the text.**

Doing Well by Talking Good:

A Topic Modelling-Assisted Discourse Study of Corporate Social Responsibility (*excerpt*)

1 The CSR-Corpus used for the present study consists of 317 CSR reports produced between 2000 and 2013 by 21 major oil companies. **2** Supplementary Appendix 3 contains a list with the names of the included companies. **3** The rationale for using these oil companies is twofold: first, these are the largest companies representing major oil-producing regions; second, they report on CSR activities consistently and make most of the reports available on their websites, which ensured good access to the data. **4** The size of the corpus is 14,806,512 tokens. **5** The data were manually collected from the websites of the companies and converted into text files. **6** It needs to be noted that for some companies, there were no CSR reports available for specific years, and hence, gaps were filled with relevant narratives taken from annual reports whenever possible. **7** Also, some companies, for example Gazprom, produced separate environmental reports and also included sections on social responsibility in the annual reports. **8** Both were included in the analysis, and hence, for some companies, we had two documents per year. **9** As our text files were converted from pdf files, there were a number of 'unwanted' characters and these were removed by using a combination of regular expressions and a Python script. **10** Because we were interested in words only, numbers and currency abbreviations were removed too. **11** To retrieve topics, we used the Mallet topic model package (McCallum 2002), which is becoming a standard topic modelling tool used in social sciences and digital humanities. **12** The Mallet package includes a stop list which contains grammatical words of English. **13** As we were primarily interested in lexical items, the stop list was used too. **14** Subsequently, the Mallet tools computed 80 topics by grouping together words and two-word combinations. **15** Subsequently, we studied all lexical items retrieved in each topic and, based on the main meanings of the items, assigned a topic label. **16** In cases in which the meaning of an item was not clear, we examined the use of the item in our corpus to detect the major senses in which it was used. **17** Each topic included, on average, 30 single lexical items and 30 two-word combinations.

Participant flow and recruitment period	
Missing data	
Adverse events	

Reporting results

This part of a research paper or dissertation is often approached by presenting and describing the findings in a systematic and meticulous manner. The researcher will identify and make comments on the themes that come out of the study when providing qualitative data. Excerpts from the raw data will frequently be used to illustrate these remarks. In textual research, this could include quotes from original sources. The findings section of quantitative research often consists of tables and figures with commentary on the data that is found to be significant. A location or summary statement, which identifies the table or figure and describes its contents, and the highlighting statements, which emphasize and describe the crucial or noteworthy facts, are common formats for this. Each table or figure needs to have a title and a number. Generally, further in-depth analysis of the findings is reserved for the Discussion section. Yet, in research articles, authors often report in-depth on their results of the findings as they are given, and the "Results and Discussion" header is frequently used to merge the two sections.

Presenting the findings of quantitative research

You'll probably be working with statistical analysis results if you did quantitative research.

The outcomes of any statistical tests you performed to compare groups or evaluate the relationships between variables should be included in your findings section. It should also indicate if each hypothesis was validated or not.

To organize the structured quantitative results logically, you should frame them based on your research questions or assumptions. For each research question and research hypothesis, you are supposed to share

- the kind of analysis that is performed (for example, a simple linear regression or a two-sample *t*-test). Your methodology section should contain a more thorough explanation of your investigation.
- a short generalization of any significant results, both positive and negative, should also be included. This can include any prominent descriptive statistics (such as *t*-scores, degrees of freedom, and *p*-values) as well as descriptive statistics (like averages and standard deviations). Keep in mind that these numbers are usually enclosed in parenthesis.
- a short declaration of how each outcome is connected to the research questions or whether the hypothesis was validated. Any results that didn't match your expectations and assumptions can be mentioned briefly; however, reserve your discussion and conclusion for any speculative analysis of their significance or importance.

Presenting the findings of qualitative research

Your findings in qualitative research may not all be immediately tied to particular hypothesis.

In this instance, you can organize the major themes or subjects that summarized from your data analysis into your Results section.

Start with broad remarks about what the data has described for each theme. You can discuss the following:

- Recurring themes of agreement or disagreement
- Patterns and trends
- Particularly noteworthy segments from the response of each individual

Next, provide clarification and citations to back up these points. Make sure you include any pertinent participant demographic data. An appendix may contain more data (such as complete transcripts, if applicable).

Task 5.3 Read the following part of the Results section about corporate social responsibility and identify whether the statements under the text are True or False. Write down T or F in the brackets behind each sentence.

Doing Well by Talking Good:
A Topic Modelling-Assisted Discourse Study of Corporate Social Responsibility (*Excerpt*)

5. RESULTS

5.1 Main topics and their distribution over time

Table 2 shows the 10 major topics ranked according to its proportion in the whole corpus. The remaining topics accounting for 2.3 per cent of the data contribute each less than 0.005 (as a ratio of a topic's alpha divided by the sum of all alphas—a low number, e.g. 0.005, indicates a very low concentration of that topic in the corpus at 0.05 per cent) and were hence not considered here. Although the main aim of CSR reports is to demonstrate company's actions and activities in relation to society and environment, it is interesting to note that still a larger proportion of the corpus focuses on financial concerns and developments, and some of the core CSR areas such as 'people, community and rights' account for just 10 per cent. Equally, 'environmental protection' and 'health and safety' constitute a smaller proportion of the whole corpus. Hence and contrary to the wider assumptions (Breeze 2013), CSR reports are not just about CSR activities; they also communicate extensively about issues related to CFP, including 'business operations', 'research and development', 'future plans and expansions', as well as 'products'. Core CSR areas identified in our corpus include 'people, community and rights', 'environmental protection', 'human capital', 'corporate governance and citizenship', 'environmental protection', and 'health and safety'. Although the two categories 'people, community and rights' and 'human capital' focus on people, they were kept separately, as each includes different groups of stakeholders. The latter contains references to primary stakeholders, that is internal stakeholders, who engage with the business directly including employees, management, shareholders, and customers, without whose participation an organization would not survive (Clarkson 1995). The latter focuses mainly on secondary stakeholders, that is people and organizations external to the companies, who do not engage with the business directly, but can influence or be influenced by it, positively or negatively. In the context of CSR, this group includes mostly local communities, media, and special interest groups (Clarkson 1995).

While the distribution of topics highlights the main themes of CSR reporting, and thus answers our first research question, we need to remember that CSR as a business field and a genre has undergone many changes. In order to understand the evolving nature of CSR practices, we analysed the topic distribution over time...

The area of CFP shows some interesting tendencies too. The most prominent CFP category is that of 'future plans/expansion', followed closely by 'research and development'...

1. The writer reports quantitative research results in the text. ( )

2. The writer reports qualitative research results in the text. ( )

3. The Results section and the Discussion section are separate in this research article. ( )

4. The writer states whether or not each hypothesis was supported. ( )

5. The writer reminds the type of analysis used in the research. ( )

Briefly interpreting or commenting on results

Interpreting and reporting the findings is the last stage of a research, the aim of which is to discover the solution to a tough question. The foundation for advancements is also provided by effectively communicating the findings. Investigators must evaluate their findings seriously and resist the need to overstate benefits or underreport harm in order to communicate in an effective manner. They have the unique ability to understand the constraints and quality of the data better than anyone else. As such, it is their duty to succinctly and clearly convey the findings together with any relevant context that may affect how the results may be interpreted. Researchers should provide this important aspect of their work enough patience, care, and attention. We consider the optimum approach for research, public health, clinical medicine, and reader interests is to adopt a "conservative" interpretation and reporting approach.

Analysing statistical analysis findings—Confidence intervals

Results are presented as a point estimate and corresponding confidence interval for both individual studies and meta-analyses. As an illustration, "The odds ratio was 0.75 with a 0.70–0.80 95% confidence interval." When comparing the experimental intervention's effect to that of the comparative intervention, the point estimate (0.75) provides the most accurate assessment of the size and direction of the effect. The range of values within which we may be fairly certain that the underlying effect truly exists is described by the confidence interval, which also conveys the inherent uncertainty in any estimate. The effect size is exactly known if the confidence interval is relatively narrow (e.g., 0.70 to 0.80). The uncertainty increases if the gap is wider (e.g., 0.60 to 0.93), but there might still be sufficient precision to decide if the intervention is beneficial. Wide ranges (e.g., 0.50 to 1.10) suggest that additional data would be required before we could reach a more conclusive judgement since we don't know enough about the effect to feel confident in the evidence.

Statistical significance and *p*-value

A statistical test's standard result, or *p*-value, represents the likelihood of finding the observed effect—or one that is larger—under a "null hypothesis." An extremely small *p*-value provides evidence against the null hypothesis by indicating that it is highly unlikely that the observed effect occurred by chance alone. It's been a standard procedure to explain whether a *p*-value is smaller than a given threshold in order to understand it. *P*-values less than 0.05, in instance, are frequently regarded as "statistically significant" and stated as being small enough to support rejecting the null hypothesis.

Task 5.4 Read the text above and answer the questions below.

1. What is the basic principle when interpreting and commenting on results?

2. What is confidence interval?

3. How to interpret and comment on confidence interval?

4. What is *p*-value?

5. How to interpret and comment on *p*-value?

Task 5.5 Read the Results section of SA6 again and answer the questions.

1. What kinds of visual devices are used in the Results section? In what way did they help the writer to present the findings?

2. How were the figure captions in the Results section written? Do they follow the same format?

3. What tenses are used in the sentences?

4. Do you think the Results section is easy to understand? How does the author make it readable?

5. What words or expressions are used to signal the information elements?

Using visuals

Graphs, pictures, and diagrams are examples of figures. A figure's objective is to visually represent intricate or striking experimental findings and analyses. Enough information should be included in the figure caption so that the reader may understand the figure without having to consult the paper's text description. To put it another way, one might understand the figure and its caption without having to read the rest of the page. There are several situations in which using a figure is appropriate:

- Using a figure strengthens your argument significantly. For instance, a graphic demonstrates the variations across treatments in a straightforward and striking manner.
- Information cannot be given as simply or plainly in other formats, such as a table or an in-text description.
- The image strengthens your argument by providing the reader with an instance or general observation from an experiment.

Figure captions

An image plus a caption are included in a full figure. When taken as a whole, these two sections need to give the reader enough knowledge to comprehend the information offered without consulting the text. To help readers comprehend a figure, a figure caption should include a succinct explanation of the image.

Because they aid readers in comprehending and accurately interpreting a figure, figure captions are an essential component of scientific data reporting. Without a suitable caption, every graph or image in a report is deemed incomplete.

Figure number: The sequence in which figures appear in a text should be used for their numbering (usually written in bold).

Figure title: A figure title explains to the audience what the graph or figure should communicate to them. This should set the scene for the figure rather than being a complete sentence or a declaration of the axis titles (usually written in bold) .

Figure description: Any information readers would require to fully comprehend the figure should be included in the figure description, which comes after the title. This usually has a normal font and should incorporate

- specific details about the statistics that are in display (average, range, etc.);
- explanations (if not already provided in a legend) for any distinct marks;
- sample sizes (*n*=X);
- and definitions of all acronyms (no need to define usual acronyms like "ml," "ppm," etc.).

All tables containing data are referenced.

DON'T just reiterate material that has already been provided in the graphic or data analysis.

Task 5.6 Read the following figure captions, discuss with a partner about the quality of their writing and identify the poor caption and the qualified caption.

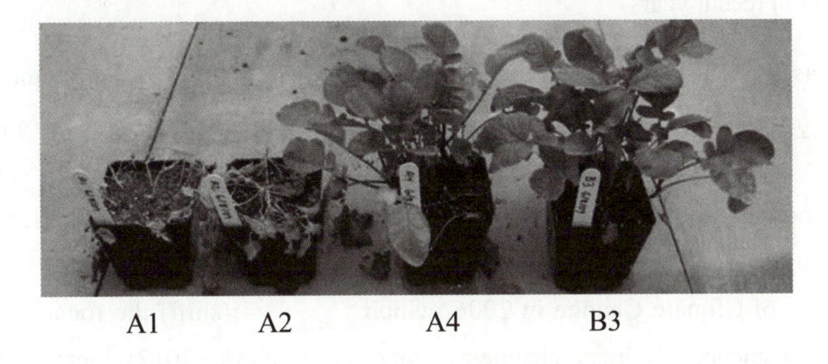

Figure 1. Radish plants subjected to a freezing treatment. It shows the effects of freezing at −15°C for 2h (A1 and A2) compared with control plants (A4 and B3) kept at room temperature. The plants in pots A1 and A4 were cold acclimated for 2 days at 2.5°C prior to freezing or room temperature treatments. The plants in pots A2 and B3 were not cold acclimated and were kept at room temperature (~ 25°C) for 2 days prior to freezing or room temperature treatments. Following the freezing or room temperature treatments, plants were kept in a greenhouse for one week.

Using appropriate tenses

It is advised to use past tenses when referring to results from hypothesis testing studies; non-textual elements should be referred using present tenses.

Task 5.7 **Read the following selections from the Results section in the Sample Articles provided and fill in the blanks with given verbs in the appropriate tense.**

1. Figure 1 ________ (show) the results in two diagrams, the first focusing on CSR and the second on CFP topics. As can ________ (see), the category 'people, community and rights' ________________________ (on the rise), which ________ (confirm) the claim that these aspects are gaining greater importance than other CSR areas.

2. Figure 3 ________ (present) the use of 'human rights' and 'climate change' in our corpus. As can ________________ (see), 'human rights' ________ (note) a steady rise accelerated after 2010, whereas 'climate change' ________ (appear) to be given less prominence, especially in recent years.

3. An increased attention to climate change after 2003 can ________ (note), and this might ________ (influence) by a number of political and media factors. The wider media campaign following the release of Al Gore's book and film might ________ (play) a role. More important from the point of businesses ________ (be) probably the ratification of the Kyoto Protocol by the EU in 2002 and the publication of the Stern Review on the Economics of Climate Change in 2006, which ________ (shift) the focus from climate change as a science to climate change as economics (Koteyko 2012). Increasingly climate change ________ (begin) to be perceived as an investment opportunity and to a lesser extent as a threat. This increased attention ________ (accompany) by a much more pro-active attitude to climate change, as ________ (reflect) in the collocational profile of 'climate change' in 2007, at the point when the term ________ (reach) a peak in our corpus. This result ________ (confirm) the tendency reported in research by Grundmann and Krishnamurthy (2010) on the media coverage of climate change, who too ________ (note) an exponential rise after 2005 and a peak in 2007.

4. As Table 3 ________ (show), in 2007, 'climate change' ________ (be) strongly associated with the action verb 'combat' and nouns pointing to goals and actions such as 'approach', 'policy', 'goal', and 'initiative'. We also ________ (find) here a number of associations that signal specific causes and preventative measures, including 'greenhouse', 'fossil', 'carbon', and 'emission'. Studying ________ (expand) concordance lines of the collocation pair 'climate change' and 'greenhouse' ________ (indicate) some of the preventative actions that the oil industry introduced or intended to introduce.

Characteristic expressions

For the Results section, you now should focus on gathering characteristic expressions. The expressions can be categorized to perform the following functions:

- Referring back to the research aims or procedures
- Explaining the data in a table or chart
- Stating the results
- Moving to the next results
- Summarizing the section on results

Scan the QR code to obtain an expression list. Make an effort to become acquainted with them and select a few to use into your own upcoming research paper composition.

Unit task

Drafting Your Results Section

In learning of this unit, you have perceived the overall structure of the Results section, and the specific devices to restate method-related details, report both positive and negative results, and interpret or comment on results afterwards. In a result, you are capable of drafting your own Results section of a research article. The importance of this section is self-evident, for it decides how the findings of your study is presented to the readers. Follow these steps to complete this task.

Step 1: Prepare the data you collected by adopting the methodology and recall your research questions.

Step 2: Compose an **OUTLINE** for the Results section.

- Present an introduction for the Results section. You can start with an opening sentence and restates the research questions.
- Report the principal findings of your research. Your findings need to be organized in consistence with the sequence of your research question.
- Illustrate your findings visually with graphs, tables, and other figures.

Step 3: Write the Results section.

Using your outline as a guide, write the Results section of your research proposal.

- Always use simple and clear language and avoid the use of uncertain or out-of-focus expressions.
- Express the findings of the study in an objective and unbiased manner. It is best to avoid over interpreting the results.
- Use sub-sections to describe the results if the research addresses more than one hypothesis. This prevents unnecessary confusion and promotes understanding.
- Include all the positive as well as the negative results which are statistically significant. Don't exclude negative results even if they do not support the research hypothesis.
- Number all tables and figures in the sequence in which they appear in the manuscript. All tables should have a descriptive caption on the head. The figures and tables should require a minimum amount of explanation in the Results or Discussion section.

Step 4: Revise your draft constantly to achieve the best results.

Check the correctness of all the graphs and figures and make sure that no values of the observations have been wrongly copied. Think about the following questions:

- Have I finished gathering my data and thoroughly examined the findings?
- Have I incorporated all the findings that are in connection with my research enquiries?
- Have I presented every finding succinctly and impartially, incorporating prominent

descriptive and inferential statistics?

- Have I indicated if each hypothesis been proven to be true or false?
- When appropriate, have I utilized tables and figures to present my findings?
- Are all figures and tables properly labeled and cited in the text?
- Is there any speculative or subjective interpretation regarding the significance of the results of my research?

Any data which have not been mentioned in the Results section cannot be discussed later. If there are too many results then try and categorize them further into subheadings.

Unit 6

Discussing Your Study

Learning objectives

In this unit, you will

- know the general function and structure of Discussion section;
- learn how to write Discussion section; and
- understand the linguistic features in writing Discussion section.

Self-evaluation

Read the Discussion and Conclusion sections of the research articles in your discipline and answer the following questions.

- What kind of information should be covered in the Discussion section?
- How do the authors organize the Discussion and Conclusion sections?
- What's the difference between the Discussion and Conclusion sections?

The Discussion section is an important part of the research paper that allows authors to showcase the study. It is used to interpret the results for readers, describe the virtues and limitations of the study and discuss the theoretical and practical implications of the research work done. It focuses on explaining and evaluating what the author found, showing how it relates to literature review and the paper, and making an argument in support of the overall conclusion.

Many authors think it is the most difficult part of a paper to get Discussion and Conclusion sections right. It is usually presented in three forms: for many papers, Discussion is the last part, in which the Conclusion is included; if a paper has a separate section for Conclusion, the Discussion may be integrated into the Result section; in some papers, Result, Discussion and Conclusion are in separate sections.

It is often easy to pull out the key elements of the Discussion section: main findings, meaning and importance of the findings, relationships with other studies, explanations, limitations and suggestions. The Discussion section of a research paper should evaluate and interpret the implications of study results with respect to the original hypotheses. It is also the place where the author can discuss the study's importance, present its strengths and limitations, and propose new directions for future research.

INFORMATION CONVENTION

Below is the Discussion and Conclusion section of Sample Article 6 (SA6). Read it and complete the table below with the numbers of sentences that perform the given functions.

History, modernity, and city branding in China: a multimodal critical discourse analysis of Xi'an's promotional videos on social media (*excerpt*)

Discussion and Conclusion

1 **1** The above analysis shows that Xi'an uses TikTok videos to construct a digitalized urban imaginary that includes attributes of both a modern metropolis and a

historical city. **2** These attributes are realized through the deployment of various linguistic and visual resources in multimodal videos. **3** In this section, we will discuss the new branding strategies and discursive features in relation to the broader socio-cultural and socio-technical contexts in contemporary China. **4** As demonstrated in the analysis of the metropolis image, Xi'an Tourism Bureau rebrands itself as a rising microcelebrity on TikTok through a series of *wanghong* rhetorical strategies, such as recommending leisure items (Stylishness), interacting with young people (Youthfulness), depicting "time travel" fantasies (Fantasy city), featuring celebrities and trending topics (Popularity), and reporting high-profile events (Internationalness). **5** These modern aspects of urban imaginary commodified Xi'an as a landscape of mass entertainment, tourism, and investment. **6** Apart from utilizing *wanghong* cultural symbols, the videos also highlight traditional cultural symbols by recreating the Tang Dynasty and reviving folk culture. **7** Through the careful manipulation of traditional symbolic images (such as Chinese calligraphy, traditional paintings, Tang poems, and "Chang'an landscape"), the historical sense of the "proto-Xi'an" as the thriving world center has come to be equated with contemporary appeals of the "pan-Xi'an" identity. **8** By resemiotizing the contemporary and traditional cultural symbols on various material artifacts, Xi'an's modern and historical urban imaginaries have become a "structured reality" (Thurlow and Jaworski 2017, 553), which discursively contribute to the development of the symbolic economy. **9** In other words, the constructed urban imaginaries created a ripple effect from online to offline engagement, attracting tourism and investment. **10** Such a visible-invisible staging of Xi'an's unique cultural images reflects what Thurlow and Jaworski (2017, 553) call the "synaesthetic rhetorics," which "toggle constantly between the material and symbolic, between the tangible and intangible."

2 **11** The unique features of Xi'an's digitalized urban imaginary are shaped by China's urban policies and the affordances of TikTok. **12** In terms of urban policies, the videos reveal the city government's exploitation of symbolic resources in branding Xi'an on social media. **13** Through highlighting the intertwined dispositions of different cultural symbols in urban policies, the city government intends to "pursue distinction" (Bourdieu 1985, 730), gaining an advantage in the current attention economy. **14** To create a distinctive symbolic system, the Xi'an government capitalizes on the *wanghong* phenomenon for its

creation of a youth-oriented urban imaginary. **15** As noted by Han (2020, 1), *wanghong* in contemporary China has "gone through a rapid process of professionalization and institutionalization," and is often represented as something more akin to consumer lifestyle choices (than real microcelebrities). **16** According to the Xinhua News Agency (2018), to follow the current *wanghong* economic model, Xi'an Tourism Bureau has adopted an entrepreneurial approach and stipulated four principles in its urban image repositioning policy, that is, utilizing the cultural narratives as a branding device, customizing Xi'an themed challenges on social media, inviting KOL (key opinion leaders) to record video logging in Xi'an, and producing personalized, glamorous short-videos related to Xi'an (Li and Jiang 2018). **17** These four principles reflect the institutional emphasis on symbolic values in rebranding the city's image. **18** By integrating the popular *wanghong* economic model into its urban policy, the Xi'an government has successfully transformed its city image from an under-developed industrial city to a trendy and popular one.

3 **19** Meanwhile, the promotional videos reflect the city government's systematic exploitation and promotion of traditional Chinese cultural symbols in its urban imaginary. **20** This can be regarded as part of the Party-state's attempt to redevelop "a comprehensive national cultural identity that integrates traditional values with contemporary life" (Cao 2014, 27). **21** As a vital part of China's soft power-building strategy, cultural rejuvenation has become increasingly highlighted in contemporary cyberspace (Wang 2017). **22** To reconnect with the glorious past of the Tang Dynasty, the Xi'an Tourism Bureau privileges the cultural symbols or "mythologies" about the ancient past of Chang'an. **23** Such a "mesmerizing new aesthetic mode of nostalgia" (Keblinska 2017, 129) can provide a "giddy escape" from reality and invoke the audience's appreciation for Xi'an's contemporary urban imaginary (Zukin et al. 1998).

4 **24** Finally, the digitalized urban imaginary is shaped by the technological affordances of TikTok, which makes it different from branding discourse in traditional media. **25** Informed by previous studies (e.g. Treem and Leonardi 2013; Feng 2019) and the present data, we summarize four central affordances of TikTok, namely editability, portability, connectivity, and multimodality. **26** Editability refers to TikTok's affordance of allowing users to create, modify, and revise their short videos, which can "allow for more purposeful communication" than traditional documents (Treem and Leonardi

2013, 160). **27** With a higher level of editorial control, Xi'an Tourism Bureau can tailor its content accordingly, with specific label choices targeting specific stakeholders (e.g. the label of "internalization" for investors, and the "revival of folk art" for tourists). **28** Portability refers to TikTok's affordance of allowing users to read the branding videos anytime and anywhere. This makes it necessary to create appealing videos and engage viewers' attention. **29** Therefore, the official videos adopt a casual expression style, draw on the currently fashionable concept of "time travel," and feature celebrities and trending topics. **30** Connectivity refers to the convenience of building connections between individuals (via the sign "@") and between an individual and a topic (via hashtags#). **31** Such connectivity can "supplement existing relationships and create a greater sense of community" (Treem and Leonardi 2013, 164). **32** Multimodality refers to TikTok's affordance of allowing users to manipulate multimodal resources such as visual depictions of characters (e.g. youngsters, celebrities, local people, and tourists) and scenes (e.g. "snowy Chang'an" and "bustling Chang'an") in TikTok.

5 **33** The socio-technological transformation of Xi'an's branding practice can shed light on the construction of urban imaginaries for emerging second-tier cities in China and other countries that are seeking transformation from a functional to a symbolic economy. **34** First, with the development of modern technologies and global consumerism, differentials in cities' "representational and promotional power have widened" (Greenberg 2000, 229), and it is important to draw on appropriate branding tactics to promote cities. **35** To distinguish themselves from highly urbanized megacities with well-established brand images, second-ranked cities need to create their unique local cultural symbols and avoid reproducing similar images (such as the use of homogeneous skyscrapers and bustling traffic to highlight economic prosperity). **36** Second, Xi'an provides a successful example of collaborating with social media platforms. **37** In the digital age, social media can be a game-changer in the urban landscape and cities need to make effective use of social media to build their urban imaginaries. **38** Third, on a more critical note, despite the benefits of symbolic economy, it should be noted that the multimodal representation of Xi'an's urban imaginary is regulated by an authoritatively constituted knowledge that limits the range of available resources (cf. Foucault 1980). **39** To some extent, it reflects Bourdieu's (2001) notion of symbolic violence, in which "the symbolic economy further constrains non-hegemonic

groups" (Schwarz 2016, 2). **40** As demonstrated in the analysis, the constructed imaginaries are always middle-class and young generation-oriented, which may make other social groups, in particular, the minority or disadvantaged groups feel unwelcomed or excluded. **41** A more inclusive approach might be needed in the creation of promotional materials to appeal to a wider range of audience.

6 **42** To conclude, this study provides a new understanding of digitalized city branding practice in the socio-cultural context of contemporary China. **43** Adopting a multimodal critical discourse analysis approach, it explicates how Xi'an's city branding discourse is shaped by the city's urban policies and the affordances of social media on the one hand, and how it is realized through the deployment of linguistic and visual resources on the other hand. **44** Xi'an's tourist-oriented urban imaginary, which is characterized by the effective hybridization of attributes of a modern metropolis and a historical city, reveals an unmistakable orientation toward symbolic economy. **45** Xi'an's successful transformation into a *wanghong* city through social media provides useful references for other cities that are striving to develop their symbolic economy. **46** Methodologically, it develops an explicit semiotic framework for systematically describing the realization of urban imaginaries through language and images in multimodal videos. **47** The framework can be applied to analyzing the multimodal construction of corporate identities, institutional identities, and so on beyond the Chinese context. **48** In the digital age where the Internet, and social media in particular, has transformed how we communicate and consume, new forms of institutional discourse are emerging rapidly. **49** Multi-disciplinary and contextualized theoretical accounts are needed to understand their new meanings and the complex semiotic resources for realizing the meanings. **50** This study is a modest step towards such an understanding and it is hoped that it can inspire further semiotic studies on various forms of digitalized institutional discourse in new contexts.

Elements	Paragraph No.
A very short introductory context that repeats the research question	
Report on data collection, recruitment, and/or participants	
A systematic description of the main findings in a logical order (generally following the order of the Methods section)	
Other important secondary findings, such as secondary outcomes or subgroup analyses	
Visual elements, such as, figures, charts, maps, tables, etc. that summarize and illustrate the findings	

General functions of Discussion section

The purpose of the Discussion section is to interpret and describe the significance of the findings in relation to what was already known about the research problem being investigated and to explain any new understanding or insights that emerged as a result of the research. The discussion will always connect to the introduction by way of the research questions or hypotheses posed and the literature reviewed, but the discussion does not simply repeat or rearrange the first part of the paper—Introduction section; the discussion clearly explains how the study advanced the reader's understanding of the research problem.

The basic functions of the Discussion section are as follows:

- To state the major findings
- To analyze data and present the relationship among data
- To expound viewpoints
- To refer to the previous research
- To point out doubts
- To state the significance
- To acknowledge the limitations
- To make suggestions
- To lead to the conclusion

Organization of Discussion section

In some disciplines, the researcher's argument determines the structure of the presentation and discussion of findings. In other disciplines, the structure follows established conventions. Therefore, it is important to investigate the conventions of your own discipline, by viewing these through articles published in your discipline and target journals. The discussion part in a research article may be

- in a combined section called "Results and Discussion";
- in a combined section called "Discussion and Conclusion"; or
- in a separate section.

You may find different arrangement of the Discussion sections in our Sample Articles.

The basic structure of Discussion section follows a specific-to-general pattern. It is often wise to write paragraphs in the editing process to reorganize the information so that it flows in a clear and logical order. Figure 5.1 provides a guideline on how to organize the Discussion section professionally. The topics discussed in this section should be organized similarly to the Results section and key elements should be separated into distinct paragraphs.

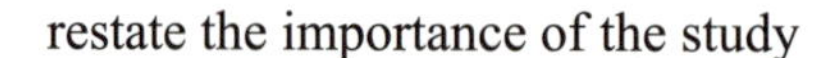

restate the importance of the study

address the research question from the Introduction section

analyze critically the major findings

relate findings to existing reviewed literature

discuss limitations of the study

discuss potential future directions of the study

Figure 5.1 The organization of the Discussion section

The results of the study need to be integrated into the Discussion section logically, not simply listed in the order they were found. The final paragraph of the Discussion section is always the conclusion and should include a brief summary of the key findings of the study along with how the findings are relevant to the field of study.

How to write a strong Discussion section

A strong Discussion section provides a great deal of analytical depth. The author should critically analyze and interpret the findings of the study and place the findings in the context of published literature and describe how the study moves the field forward. It is often easy to organize the key elements of the Discussion section into distinct parts.

Part 1: Remind readers the importance of the study.

In the first few sentences of the Discussion section, state the main problem that you were trying to address. Although it should be relevant to the information that you provided in the Introduction section, this part should not repeat statements that have already been made. After identifying the problem, state the main reason why this study is necessary. Describe how answering this specific research question will make a significant contribution to your field.

In the following example, the problem as well as the significance and the ultimate reason for performing the study are presented:

Main conflict of the landscape architecture heritage protection lies in weak awareness of value of heritage resources, excessive development of heritage sites, and vacancy in management and monitoring of heritage resources. Management mode of many world heritage sites still follows traditional mechanism.

Part 2: Provide a critical analysis of your major findings.

In one or two sentences, state the main methods that you used to study the specific research questions. The focus of this part is to highlight the most important contribution that your study has made. Explicitly state this result. Additional findings can be described in subsequent paragraphs. Do not repeat detailed results that can be found in the Results section. Multiple paragraphs may be needed depending on the number of key studies that exist on your topic. This part should be well-rounded, meaning that contrary reports must also be discussed. In the case of a contrary report, you should state your interpretation of how and why the results of the two studies differed.

In the following example, the approach and the main result is stated:
Additional Census details have shown that ownership of desktop, laptop, and tablet devices has declined, while smart phone ownership has increased.

Part 3: Discuss additional findings and how these fit with existing literature.
Most studies yield multiple results. Unexpected and interesting findings may be especially important to convey to readers. In addition, if a finding is contrary to what has been suggested in the literature, acknowledge this, and offer explanations based on your study. Even if a result was not statistically significant, it can be also helpful to discuss a potential trend that may be important to assess in a future study. If these additional findings relate to your main finding, discuss the associations. For example:
Researchers found that 25% of Hispanics, 23% of Blacks, and 13% of Whites lacked home broadband but owned a smartphone (Perrin & Turner, 2019).

Part 4: Discuss the limitations of the study.
Discuss potential limitations in study design. For example, how representative was your approach? Did sample size affect your conclusions? Consider how these limitations affect the interpretation or quality of data. Do they affect the ability to generalize your findings?
However, interaction during the pandemic may have prioritized instructor student interaction over peer interaction, perhaps in part in order to prioritize student flexibility and access (Rutherford et al., 2021).

Part 5: Discuss future directions.
Most studies yield new discoveries that prompt additional studies. Consider what new directions are supported by your findings. Making recommendations for follow-up studies is an important part of the Discussion section. For example:

Research in this area should focus on replicable, scalable interventions that online instructors can adopt to increase interaction at both the instructor and peer levels, particularly at the K-12 level.

Part 6: Discuss your overall conclusion and the major impact of your study.

Relate this section to the first paragraph of the Discussion section. In other words, how your study address the issues that you presented in the Introduction section and re-stated earlier in paragraph 1 of the Discussion section. For example:

With an eye on the physical, human, and social resources students at all levels of education need, the U.S. education system can not only improve online learning but also positively affect the broader educational environment, which with growing frequency makes use of online resources, and thus help the nation tackle educational underachievement and inequity.

In brief, a strong Discussion section includes a concise summary of the problem you are investigating and a critical discussion of major and minor findings in the context of published literature. The limitations should also be acknowledged, and future directions should be supported. A strong ending is important to discuss the significance, overall conclusion, and major impact of your study.

Some strategies can help to draft a good Discussion section:

- Explain how your findings/results relate to what has been already known in the field as well as what you expected to find. You should refer back to your Introduction section and determine if what you found were consistent with the existing literature, or if they were somewhat unexpected or controversial.
- If your findings were unexpected and/or contradictory, you need to explain why that was the case. Did your sampling method contribute to it? Or your choice of methodology? At this point, make sure you have sufficiently justified your methodological decisions in the methodology part of your thesis. Unusual findings can be good, but they might also elicit more questions from other readers, so make sure you have all the answers.
- Try to show both sides of your argument. Be your own devil's advocate. This will give your conclusions more credence.

- Again, somewhere in your Discussion section show that you are aware of the limitations of your study.
- Provide one or two recommendations for future research or follow-up studies.
- Make sure all your results have been addressed, including those that were not statistically significant.
- You might revisit your Introduction section at this point and put more emphasis on studies that have proven relevant for the interpretation of your results.

There are some similarities between the terms “discussion” and “conclusion” in academic writing. These two terms usually represent two separate concepts. While you might have noticed some similarities between these two, they also generally have different purposes. The discussion is a **detailed presentation** of your findings and provides scientific back-up for your arguments. It **explains** your findings and **interprets** them in the context of previous work, as well as providing some suggestions for future research. The conclusion, on the other hand, is generally **brief** and provides just the **main points of your study**. It can be seen as a summary of your discussion and tells the reader why your research matters. It should have a clear structure: a beginning (introduction), a middle part (synthesis of your findings), implications for theory and practice, and an end (future directions). Conclusions should be able to stand on their own; they are an entity by themselves. The author need to conclude with a short paragraph and use their own words for this final statement, and it might be the best to avoid direct quotes in the final pages.

Task 6.1 Assess the veracity of the following statements as either True or False. Write down T or F in the brackets behind each sentence.

1. Draw appropriate conclusions based on the data without overstating the findings. ( )
2. Placing excessive emphasis on the study’s limitations potentially leads readers to question the study’s relevance. ( )
3. Recognize and address the limitations of ther study. ( )

4. Repeat the introductory content without effectively tying it to the results. ( )

5. Omit any form of conclusion. ( )

6. Introduce the themes that were already covered in the study's results and finding. ( )

Task 6.2 Choose two research papers in your discipline, read the Discussion and Conclusion sections, and compare their structure and content.

Linguistic features

Discussing research findings and knowledge in academic contexts, particularly in the written mode, has always seemed to be a technically challenging task. The arduous publication process aside might be due in part to the complicated nature of the academic content, but also the linguistic demands of scientific writing. As for the linguistic dimension, various components should be taken into account in reporting the study which range from the use of academic lexicon and formal tone to the use of discourse markers such as cohesive devices and hedges. These devices can play important roles in conveying the message as well as transmitting the author's degree of certainty about the message.

Tense

The Results section was completed before the paper is written. Therefore, the simple past tense is the natural choice when describing the results obtained. In the Discussion section, the past tense is generally used to summarize the findings. But when you are interpreting the results or describing the significance of the findings, the present tense should be used. Therefore, a combination of the past and the present tense is usually used in the Discussion section.You may also need to use the future tense in the Discussion section if you are making recommendations for further research or providing future directions. For examples:

Management mode of many world heritage sites still follows traditional mechanism. (present tense)

As a theoretical paper aimed primarily at a research audience, my focus has largely been on what the literature suggests about the phenomenon that is teacher curriculum supplementation, distilled in the large, leftmost box of Figure 1. (past tense, perfect tense)

As a research community, we must develop our understanding of teacher curriculum supplementation if we are to explore it as a potentially powerful lever for improving instruction. (future tense)

Hedges

Hedging language makes our sentences more accurate and truthful. We can use them to avoid generalizations, inaccuracies, and stereotypes. Each of these could prove very problematic in academic writing.

The following sentences are modified in hedging language:

Many *old people are not good with technology.*

A lot of *old people are not good with technology.*

Old people ***often*** *struggle with technology.*

It is sometimes said that *old people are not good with technology.*

Old people ***sometimes*** *struggle with technology.*

Some people think that *old people are not good with technology.*

It could be argued that *most old people are not good with technology.*

Hedges are linguistic devices that control the degree of fuzziness in communicating messages, helping the authors express how certain they are about the true value of their statements. Hedging strategies used in the Results and Discussion sections of research articles can be very illuminating as these sections play a key role in communicating and interpreting the research findings. Some researchers used Hyland's taxonomy of hedges for study and found that the Discussion sections of qualitative articles are more heavily hedged than the Discussion sections of quantitative articles.

Having considered both formal and functional criteria for identification of hedges, Salager-Meyer offered a taxonomy of linguistic devices through which hedging can be expressed:

- **Modal auxiliary verbs**: *may, might, can, could, would, should*
- **Modal lexical verbs**: *to believe, to assume, to suggest, to estimate, to tend, to think, to argue, to indicate, to propose, to speculate*
- **Adjectivalmodal phrases**: *possible, probable, unlikely/likely*
- **Adverbial modal phrases**: *perhaps, possibly, probably, practically, likely, presumably, virtually, apparently*
- **Nominal modal phrases**: *assumption, claim, possibility, estimate, suggestion*
- **Approximators of degree, quantity, frequency and time**: *approximately, roughly, about, often, occasionally, generally, usually, somewhat, somehow, a lot of*
- **Introductory phrases**: *I believe, to our knowledge, it is our view that, we feel that*
- **If clauses**: *if true, if anything*

Examples can also be found in our Sample Articles:

It is worth noting that the absolute of the coeffcients of ideological variables in Panel D is larger than those in Table 2 (full samples), ***indicating*** *that after we exclude the extreme ruling party in our samples, the influence of political party on innovation is more significant and larger compared to our earlier finding.*

Moving forward, our understanding of teacher curriculum supplementation ***would be*** *enhanced by studies of ...*

Evidence ***suggests*** *that intentionally increasing interaction improves student learning.*

However, studies that focus on a cognitive perspective ***often*** *look at the discursive as means to comprehend the cognitive self.*

Characteristic expressions

There are mainly seven types of characteristic expressions for the Discussion (and Conclusion) section of a research article:

- Summarizing the study results
- Linking findings to previous research
- Discussing research results
- Discussing limitations
- Discussing implications
- Directions for future research
- Closing statement or paragraph

Using these characteristic expressions can help structure the Discussion (and Conclusion) section effectively, clarify the significance of the findings, and guide readers through the implications of the research. Scan the QR code to acquire more expressions for different purposes in this section.

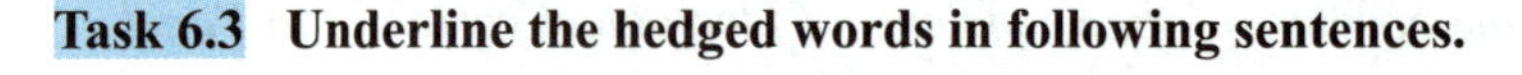

Task 6.3 Underline the hedged words in following sentences.

1. Evidence seems to suggest that juggling work, studies and family commitments contributes to stress among students.

2. Apparently, it rains in England all the time. That's why everywhere is so green.

3. That could be the reason why these students go abroad to study.

4. There's a strong possibility that this new technique will be very successful.

5. The reason she cannot sing is undoubtedly because she has never been taught.

Check your understanding

Task 6.4 Pick two journal articles in your field. Analyze the organization and expression of the Discussion sections.

Unit task

Drafting Your Discussion Section

After your study is carried out, it is time to presenting your own Discussion section on your selected research topic. Do the following to finish this task.

1. Summarize the key points.

2. Analyze your research before relating how your research fits into the field as a whole.

3. Compare your work to the gap in the field, including how your research might have moved the edge of current knowledge.

4. Describe how your research modified our view of what lies beyond the edge of current knowledge.

5. Provide some suggestions for future directions on how to examine those hypotheses are needed.

Unit 7

Writing the Title and Abstract

Learning objectives

In this unit, you will

- understand the general function and purposes of the title and the abstract;
- learn about the common information elements in the abstract; and
- develop linguistic strategies for writing effective titles and abstracts.

Self-evaluation

How would you write the title and the abstract of your paper? Read the titles and the abstracts of two articles in your research field and answer the following questions.

- How long are the article titles?
- Are the titles written in full sentences or phrases?
- What kinds of information are given in each title?
- How long is the abstract of each article?
- Do the abstracts have a single paragraph or multiple sections?
- What kinds of information can you identify in the abstracts?

Writing an original research article can be very complicated. After having done so much research—reading so many papers and doing so many analyses on your data—it is easy to get caught up in the details and lose sight of the big picture. A good manuscript is more than the sum of its parts. Manuscripts should be built around a unifying message. The earlier you decide on this message, the easier it will be to create a harmonious whole. This is not to say that you need to have a refined version of your message at your fingertips when you first sit down to write. On the contrary, your message will evolve as you write and rewrite. Working through successive drafts will bring new insights into your work and help you discover new relationships among your data and analyses. Once you know what you want to say, you can concentrate on how to say it clearly, correctly, concisely, and convincingly. This is why the conclusion, the title, and the abstract should be the last part to be written and added to the manuscript. Admittedly, title and abstract in published papers are at the same time both front matter and summary matter.

This unit will help you understand how to write a concise and attention-drawing title and an effective abstract for your paper.

WRITING THE TITLE

Below are the titles of some research papers. Discuss with your partner about the structure of each title and the information elements given in each title.

1. Organization and Evolution of Climate Responsive Strategies, Used in Turpan Vernacular Buildings in Arid Region of China

2. Theories and Methods of Landscape Architecture Heritage Protection

3. Contrasting Suitability and Ambition in Regional Carbon Mitigation

4. Good Deeds Done in Silence: Stakeholder Management and Quiet Giving by Chinese Firms

5. Does the Shale Gas Boom Change the Natural Gas Price-Production Relationship? Evidence from the U.S. Market

6. Journalistic Construction of Congruence: Chinese Media's Representation of Common but Differentiated Responsibilities in Environmental Protection

7. History, Modernity, and City Branding in China: a Multimodal Critical Discourse Analysis of Xi'an's Promotional Videos on Social Media

Functions of the title

The title is arguably one of the most important parts of a research paper. It tells the gist of the study and is always the first or even the only part that people read about the paper. Prospective readers search for key words in search engines that return lists of titles linked to abstracts. Metaphorically, the title is the "face" of your paper. It contains the first words the readers will see and gives them a first impression of how well your paper may satisfy their needs. Your title is unique, which identifies your paper in references and databases. What makes your title unique is the way its keywords are assembled to differentiate your work from the work of others.

Academically the primary function of a title is to provide a precise summary of the paper's content. A title should stand alone and be fully explanatory without further elaboration. A reader browsing through paper titles in an online database should be able to quickly read your title and know exactly what your paper is about. In brief, a title is supposed to fulfill multiple purposes for the reader and the writer.

For the reader, a title

- helps decide whether the paper is worth reading further;
- gives a first idea of the contribution (a new method, chemical, reaction, application, preparation, compound, mechanism, process, algorithm, or system);

- provides clues on the paper's purpose (a review, an introductory paper, etc.), its specificity (narrow or broad), its theoretical level, its nature (simulation, experimental, etc.), and the knowledge depth required to benefit from the paper; and
- informs on the scope of the research, and possibly on the impact of the contribution.

For the writer, a title

- gives room to keywords for search engines to find the work;
- catches the attention of the reader;
- states the contribution in a concise manner;
- differentiates the title from other titles; and
- attracts the targeted readers and filters out the un-targeted ones.

Task 7.1 Read the paper titles below. Based on the above discussion, analyze how the titles fulfill the purposes from the perspective of the reader and the writer.

1. From researchers to academic entrepreneurs: a diachronic analysis of the visual representation of academics in university annual reports
2. *Positive Energy* Douyin: constructing "playful patriotism" in a Chinese short-video application
3. Identity performance and self-branding in social commerce: A multimodal content analysis of Chinese *wanghong* women's video-sharing practice on TikTok
4. Design-based research: What it is and why it matters to studying online learning
5. Critiquing Empire Through Desirability: A Review of 40 Years of Filipinx Americans in Education Research, 1980 to 2020

What is a good title?

A consensus on research paper titles is that the title should be the fewest possible words that adequately describe the content of the paper. The ideal title length is about 12–15 words. The title should interest and engage the reader, use active words, and accurately reflect the study. A synthetic, informative, and crisp title captures the attention of readers and affects how your paper will be found in search engines, while an unsuccessful one will discourage readers. In other words, the job of a title is not simply presentational; a well-crafted title expands the reach of the work. In brief, a successful title should be specific, as short as possible, straightforward, and not sensational. The following three points are suggested as of what a good title is required to be.

- The title should contain key words used in the manuscript and define the nature of the study.
- The title could result in greater clarity of the message and greater attractiveness for the readers. The articles with "good" titles would then receive more attention and be used more often, and thus receive more citations.
- The title should describe the content of the paper with the less possible words. The objective, subject and result should be included. It needs to show the main idea with the least number of words to optimize the search engine results.

To sum up, a good title can be formulated briefly and sharply, being precise, simple, and short. The best title is one that gives the most accurate information about the content of the paper with the fewest possible words.

Task 7.2 Compare the following pairs of titles. Which title is better and why?

1. Heritage Protection Theories and Methods of Landscape Architecture
   Theories and Methods of Landscape Architecture Heritage Protection

2. Job Insecurity on Psychological Well- and Ill-Being among High Performance Coaches
   Impact of Job Insecurity on Psychological Well- and Ill-Being among High Performance Coaches

3. Organization and Evolution of Climate Responsive Strategies, Which are Used in Turpan Vernacular Buildings in Arid Region of China

   Organization and Evolution of Climate Responsive Strategies, Used in Turpan Vernacular Buildings in Arid Region of China

4. A Multimodal Content Analysis of Chinese *wanghong* Women's Video-Sharing Practice on TikTok

   Identity Performance and Self-branding in Social Commerce: A Multimodal Content Analysis of Chinese *wanghong* Women's Video-Sharing Practice on TikTok

5. Suitability and Ambition in Regional Carbon Mitigation

   Contrasting Suitability and Ambition in Regional Carbon Mitigation

Structure of the title

Titles can be classified as indicative or informative. The former indicates what the article is about (e.g., the relationship between A and B), while the latter informs the reader of what the study has found (e.g., A increases B). Different linguistic strategies, such as omission of articles, nominal group constructions, simple full-sentence constructions, and monosyllabic verbs and/or nouns in titles, contribute to constructing informativity in the sense that the topic, which will be further discussed in the paper, is presented in miniature in the title. Whichever type the titles are, they often contain some basic information elements:

- the **topic**, i.e., the main, general subject you are writing about;
- the **focus**, i.e., a detailed narrowing down of the topic into the particular area of the research.

The grammatical structure of titles roughly falls into four categories: nominal group (nominal phrases), compound construction, full sentence (declarative sentence), and question construction. The question construction is generally more appropriate for titles of editorials, commentaries, and opinion pieces.

Nominal-group (nominal phrases) construction

Nominal-group construction is the most commonly observed structure in the titles of research articles. Of this type, three syntactic structures are identified and respectively named as uni-head, bi-head, and multi-head nominal group. Specifically, the uni-head construction is a noun or noun group followed by a post-modifier or a prepositional phrase. The bi-head construction contains two nominal groups followed by post-modifiers while the multi-head construction contains more than two nominal groups. Below are some examples:

The Impact of Leadership on Trust, Knowledge Management, and Organizational Performance

The Role of Negative Information in Distributional Semantic Learning

A Generalized Rightward Movement Analysis of Antecedent-contained Deletion

*Concentration and Persistence of Health Spending: Micro-[KG-*5] Empirical Evidence from China*

Market Discipline and EU Corporate Governance Reform in the Banking Sector: Merits, Fallacies, and Cognitive Boundaries

Business Intelligence, Predictive Analytics and Management Accounting: A Field Study

Ecology, Community and Delight: Sources of Values in Landscape Architecture

Compound construction

Compound construction is also common among research paper titles, with a high disciplinary prevalence in the social sciences. The compound titles normally have two noun phrases (or non-clausal small independent grammatical units) which are juxtaposed on either side of (most usually) a colon or a dash. This type of titles evidences an interrelationship between the two parts constituting them, thus succinctly showing the presentation of the object of study in two different ways: problem–solution, general–specific, topic–method, and major–minor. The first noun phrase generally functions as the main title and should stand alone, while the one after the colon/dash functions as the subtitle which complements the main title by providing supplementary information. In this sense, this type of construction shows that titles are not only a succinct presentation of a given study but also a succinct reference to a specificity related to that study, evidencing a sort of cadence from the general to the particular. See the examples below:

Maps of Bounded Rationality: Psychology for Behavioral Economics

Management Ownership and Market Valuation: An Empirical Analysis

Cotext as Context: Vague Answers in Court

Numerical and Arithmetical Cognition: A Longitudinal Study of Process and Concept Deficits in Children with Learning Disability

Regional Design: Recovering a Great Landscape Architecture and Urban Planning Tradition

Emotions: From Neuropsychology to Functional Imaging

Equilibrium in Competitive Insurance Markets: An Essay on the Economics of Imperfect Information

Full-sentence construction

Full-sentence title is more frequently observed in science papers, and is also indicative of a marked contrast with respect to the same construction in the titles of the social sciences. Such titles usually express the study results or observations. Writers may adopt this structure to highlight their findings and show their confidence of their work. See the examples below:

The Consumption Response to Minimum Wage Increases

Learning Induces a CDC2-related Protein Kinase

Question construction

Question construction title reveals in general a very low occurrence, but it still has a higher occurrence in the review papers (RVP) of the social sciences than the titles of the biological sciences. The question construction in this genre seems to allow authors the possibility of posing questions on such object as an indication that, in spite of the current state-of-the-art about it, there are, still, queries in need of reply, interpretation, and conclusion. In this sense, question titles parallel science as a question process. See the examples below:

Does the Flynn Effect Affect IQ Scores of Students Classified as LD?

Do Investment-Cash Flow Sensitivities Provide Useful Measures of Financing Constraints?

Can the Use of Cannibis (Hemp), When Mixed with Additives, Be a Suitable Substitute to Conventional Building Materials?

Is Academic Writing Becoming More Informal?

Is Ability Grouping Beneficial or Detrimental to Japanese ESP Students' English Language Proficiency Development?

As frequently detected, titles in the social sciences are usually shorter than those in the biological sciences. And titles in linguistics are the shortest in the group of the social sciences, a peculiarity which is even more marked with respect to the length of biochemistry, biology and medicine titles. In view of the above-mentioned types, social sciences show a higher degree of flexibility for title formatting in contrast to the other scientific disciplines. These types of titles evidence a preference to the straightforward presentation of the object of study.

Task 7.3 **Read the following pairs of titles and explain how the two versions are different.**

1	*Version 1*	The Analysis of the Effects of the Use of Computers in Njombe Region
	Version 2	The Use of Computers: An Analysis of Their effects within Njombe Region
2	*Version 1*	Financial Poverty and the Life of Primary School Teachers in Tanzania: An Analysis of Its Effects on Their Teaching Work
	Version 2	An Analytic Study of the Effects of Financial Poverty and the Life of Primary School Teachers in Tanzania and Their Teaching Work
3	*Version 1*	Assessing the Role of Civil Laws in the Question of Marriage
	Version 2	The Assessment of the Role of Civil Laws in Marriage
4	*Version 1*	Analysis of the Impact of Agricultural Extension on Farmer Nutrient Management Behavior in Chinese Rice Production from the Perspective of Household
	Version 2	The Impact of Agricultural Extension on Farmer Nutrient Management Behavior in Chinese Rice Production: A Household-Level Analysis
5	*Version 1*	On the Role of Research-Led Teaching in the Development of Performing and Creative Arts
	Version 2	A Reflective Perspective on the Challenges Facing Research-Led Teaching in the Performing and Creative Arts

Task 7.4 Read the titles of the following articles and complete the table below.

Title	Is the title a noun phrase, a declarative sentence, or a question?	How many words are there in the title?	Can the title be shortened?
A theoretical framework for studying teachers' curriculum supplementation			
Equity in online learning			
Pronunciation in the United States is more different than in the United Kingdom			
Aesthetics, technology, and social harmony: constructing a "green China" image through eco-documentaries			
The impacts of government ideology on innovation:What are the main implications?			

How to write an effective title?

Generally, the title should indicate answers to the important basic questions: **What? Where? How?** To write a good title, think about the following questions; the answers to these questions will help you figure out what to include in your title.

- What is my paper about?
- What methods/techniques did I use to perform my study?
- What or who was the subject of my study?
- Where was the study carried out, in the laboratory, or in the field?
- How was the study organism or phenomenon examined?
- What were the results of my study?

When writing the title, keep the following tips in your mind:

- Think about terms that people would use to search for your study and include them in your title.
- Pick keywords from recent or often-cited titles close to your work. Use clear and specific keywords.
- Use appropriate descriptive words.
- Choose strategically to write your title into a noun phrase or a full sentence so that the information to stress appears near the front.
- Avoid having too many details. Keep the title brief and attractive without omitting key information.
- Delete unnecessary and redundant words to meet a suitable word count. Phrases such as "role of," "effect of," "use of" and "report of a case of" can often be omitted.
- Shift some words around and rephrase the title to make it sound more natural and exact.

Subtitles are quite common in social science research papers. The subtitles may help to

- provide additional context or information;
- add substance to a literary, provocative, or imaginative title;
- qualify the (geographic/temporal) scope of the research;
- and focus on investigating the ideas, theories, or work of a particular individual.

Task 7.5 How are the titles in each pair different from each other? Which one do you like better? Why?

Pair 1

T1: Too credulous to be trusted: The gullible Lord Byron

T2: Was Lord Byron too credulous to be trusted?

Pair 2

T1: An empirical study on the evaluation of course ideology and politics in University Mathematics

T2: On the evaluation of course ideology and politics in University Mathematics: an empirical study

Pair 3

T1: Learning commons (designing library space): A comprehensive working model for academic libraries

T2: A comprehensive study of the working model for academic libraries of learning commons (designing library space)

Pair 4

T1: Gender discrimination in healthcare expenditure: A study of Indian states

T2: A gender discrimination study in the healthcare expenditure of Indian states

Pair 5

T1: Exploring the path of labor beauty of primary and secondary school students in the post epidemic era

T2: On the study of the path of labor beauty of primary and secondary school students after the epidemic

Pair 6

T1: Innovative measures in enhancing educational research in teacher education

T2: On the innovative measures taken to enhance the educational research in teacher education

Pair 7

T1: Work complexity of urban cleaning professionals in the city of São Luís de Montes Belos, Goiás, Brazil: Street job conditions and variability

T2: Work complexity of urban cleaning professionals in the city of São Luís de Montes Belos, Goiás, Brazil

Pair 8

T1: Oral English teaching in middle schools guided by constructivist learning theory

T2: How to teach oral English in middle schools: from the perspective of constructivist learning theory

WRITING THE ABSTRACT

Read the abstract of Sample Article 1 (SA1) and complete the table under the text. Recall your memory of the information chunks and elements in different sections of research articles. Decide which chunk or element each sentence in the abstract represents. The first sentence has been given as an example.

Testing the Relationship of Linguistic Complexity to Second Language Learners' Comparative Judgment on Text Difficulty

ABSTRACT

1 This study examined the relationship of linguistic complexity, captured using a set of lexical richness, syntactic complexity, and discoursal complexity indices, to second language (L2) learners' perception of text difficulty, captured using L2 raters' comparative judgment on text comprehensibility and reading speed. **2** Testing materials were 180 texts abridged from college English coursebooks, and raters were 90 advanced Chinese learners of L2 English. **3** Forty-five raters read paired texts and determined which text was harder to understand in each pair, and another 45 raters read paired texts and determined which text they read faster in each pair. **4** Two stepwise linear regression models containing lexical, syntactic, and discoursal features explained 48.1% and 54.6% of the variance in L2 learners' estimates of text comprehensibility and reading speed, respectively, outperforming four commonly used language readability models. **5** These findings contribute useful insights into the relationship between linguistic complexity and L2 learners' perception of text difficulty.

lexical 词汇的
syntactic 句法的
discoursal 语篇的
indices（名词 index 的复数）指数，指标
perception 认知；理解力
abridge 删节
stepwise 渐进的；阶梯的
linear regression 线性回归
variance 变化；方差
outperform 优于；超额完成
contribute 增加；增进

Sentence	Information chunks or elements
1	Outlining purpose of the research
2	
3	
4	
5	

Structure of the abstract

The abstract is a synopsis of the paper, a summary containing a brief account of the content, purpose and theoretical background of the study, which is accessible via online indexes and abstract databases. Thus, abstracts function as independent discourses (Van Dijk, 1980) as well as being advance indicators of the content and structure of the following text. Along with the title of the research article, the abstract helps readers decide whether the paper is related to their research interests. It condenses a longer piece of writing while highlighting its major points. Routinely, the abstract is composed after the paper is completed. The ideal length for this section is between 100 and 150 words. Think of the abstract as a way of "selling" your research, particularly as most research archives contain abstracts so others can decide whether to read them (Parrott, 1999). You need to keep balance between conciseness and clearness. It is suggested that only economical words be used to state clearly what the paper is about and only essential information be provided for readers.

Types of abstracts

Abstracts fall into two major categories: indicative and informative.

An indicative abstract is usually a single paragraph telling readers what to expect if they read the article; it informs the readers of what the writer will deal with or attempt to prove, rather than a synopsis of the actual results. It helps readers to understand the focus, arguments and conclusions of the larger document so that they can determine whether to read it more thoroughly. This type of abstract is more appropriate for review articles or case reports.

In contrast, an informative abstract is a condensed version of the article. It is used for more strictly structured documents (like scientific experiments or investigations) and includes the elements of the original research report: its **objective**, **methods**, **results**, and **conclusions**. It gives a summary of the main factual information, such as the materials and methods, the results and conclusions, etc. This type of abstract is more suited to reports of original research works. It should be written to stand alone; in other words, readers should be able to understand the abstract without reading the entire article. Normally, when writing up research, the informative abstract is better since you give the reader factual information as well as your main opinions.

Formats of abstracts

Two different formats are often seen in abstracts of academic journal articles. A simple abstract is an undifferentiated paragraph, while a structured abstract contains typical sub-sections as follows:

- **Background**: one or two sentences
- **Aim (purpose or objective)**: one or two sentences
- **Methods**: two or three sentences
- **Results**: no more than ten sentences
- **Conclusion** (contribution, implications, etc.): one or more sentences

Journals often have their own preferences or requirements in abstract writing. You should read your target journal to find out how to write the abstract of your paper.

Information elements

As a miniature of the article, the abstract would include details that allow readers to grasp the main points of the article. It may serve as an "outline" manual. Figure 7.1 presents the common information elements (IEs) in the abstract.

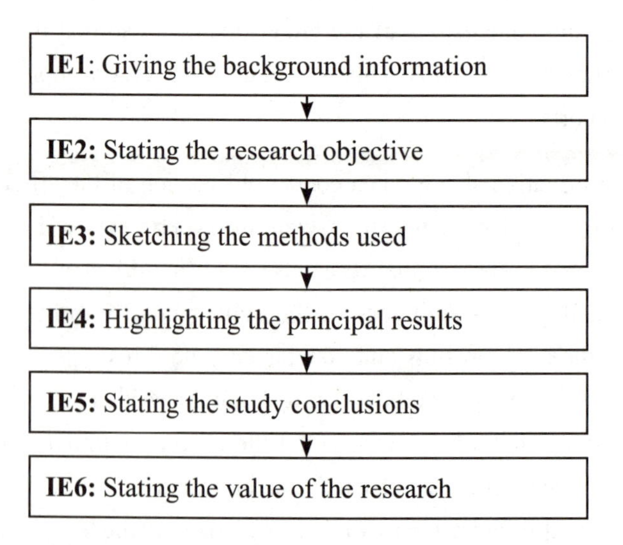

Figure 7.1 Information elements in the abstract

Not all the elements may appear in an abstract. Some authors may start right from declaring the study objective without mentioning any background information. It is also common that some elements are integrated. For example, the problem concerned and the aim of the study may be presented in one sentence, or the methods and results are described together. Generally, the elements in an abstract are arranged exactly in the order shown in Figure 7.1, through which the abstract reflects the essential components as well as the organization of the paper.

Task 7.6 Below is the abstract of Sample Article 6 (SA6). Read the text and identify the "Background," "Objective," "Methods," "Results," and "Conclusion." Complete the table under the text and state your reasons.

1 In the digital age, cities around the world are mobilizing various symbolic resources to rebrand their images through social media. **2** Against this background, this study investigates how Xi'an, a second-tier developing city in China, constructs its digitalized urban imaginary using the popular social media platform of TikTok. **3** A semiotic framework is developed to model Xi'an's urban imaginary as evaluative attributes and to elucidate how they are constructed through linguistic and visual resources in short videos on TikTok. **4** The analysis of 294 videos shows that Xi'an highlights its dual identity as a modern metropolis and a historical city. **5** The modern metropolis image is characterized by the personification of Xi'an as a stylish, young, popular, and international microcelebrity; the historical city image is constructed through recreating the Great Tang Dynasty and revitalizing local folk art. **6** The characteristics of city branding discourse reflect China's *wanghong* economy, urban policies and the affordances of social media.

Background	
Objective	
Methods	
Results	
Conclusion	

How to write the abstract

Writing a good abstract requires that you explain what you did and found in simple, direct language so that readers can grasp the essence of your work and decide whether to read the whole piece of writing for details. You may write your abstract by the following structure:

- Give a basic introduction to your research area, which can be understood by researchers in any discipline. (1–2 sentences)
- Provide more detailed background for researchers in your field. (1–2 sentences)
- Clearly state your main result. (1 sentence)
- Explain what your main result reveals and/or adds when compared to the current literature. (2–3 sentences)
- Put your results into a more general context and explain the implications. (1–2 sentences)

A good abstract is usually short, never exceeding the word limit as required by the journal. It should follow the 4Cs principle:

- **Complete**: covering all the major parts of the research work
- **Cohesive**: flowing smoothly throughout
- **Concise**: containing no extra words or unnecessary information
- **Clear**: remaining readable to both experts and non-experts, even in its condensed form

To write a good abstract, you may need to decide the key messages the paper tries to deliver. Restrict the study results to those that explain key messages, and include methodology that enables the reader to know how those results were obtained. Below are some specific guidelines to consider in writing abstracts.

- Consult the journal's instructions for authors for special requirements.
- Do not begin the abstract by repeating the title. Do not cite references.
- Provide absolute results for main outcome measures (e.g., report incidence rates rather than reporting only relative risks). Provide confidence intervals whenever possible (if not, provide *p*-values).
- Include key terms and describe databases and study groups (related to the subject under discussion) so as to facilitate the abstract being searched in retrieval systems.
- Ensure that all concepts and data in the abstract are consistent with those in the text.

Task 7.7 The following sentences are the disordered abstract of Sample Article 5 (SA5). Please rearrange the sentences to produce a good abstract. Then you may check up your text by referring to the abstract of SA5.

1. Factors affecting EFL students' participation may differ by student and lecturer perceptions, which will cause differences in these stakeholders' solutions to classroom difficulties.

2. English as a foreign language (EFL) students' lack of engagement directly affects their learning outcomes, and it has captured the attention of many researchers.

3. The study further employed semi-structured interviews with six students and four lecturers.

4. Accordingly, this study measured non-English major (NEM) students and EFL lecturers' perceptions of factors hindering students' participation in English-speaking classes.

5. On the one hand, the students did not consider the large class size, insufficient time for in-class practice, and students' tendency to remain silent as hindering their class participation.

6. About 156 NEM freshmen and 14 lecturers responded to a 35-item questionnaire containing five primary clusters: linguistic, cognitive, affective, pedagogical, and social-cultural factors.

7. The results found some significant differences between lecturer and student perceptions.

8. Based on the findings, teacher professional development activities are expected to be one of the solutions to EFL students' insufficient engagement in English-speaking classes.

9. On the other hand, the lecturers considered vital barriers: students' insufficient proficiency, teachers' poor lessons, or teacher-student relationships.

The proper order: ______ ______ ______ ______ ______ ______ ______ ______ ______

Check your understanding

Task 7.10 Read the following abstract of Sample Article 4 (SA4) and discuss the questions below.

Abstract

1 This paper studies the convergence of environmental sustainability and its main determinants in selected American countries. **2** In addition, it studies the impact of economic activity, income inequality, trade openness, and innovative activity on the sustainability of these countries. **3** Convergence tests such as unit root and club convergence are applied. **4** Furthermore, cointegration and causality tests are used, and long-term parameters are estimated using methods robust for cross-sectional dependence. **5** The results show evidence of stochastic convergence with the univariate unit root tests in the five indicators (energy consumption, carbon dioxide emissions, ecological footprint, energy intensity, and load capacity factor) used, while with the panel data unit root tests only in four (carbon dioxide emissions, ecological footprint, energy intensity, and load capacity factor). **6** There is no evidence of convergence towards a single club considering the complete sample, but there is evidence of convergence towards several clubs. **7** The variables are integrated of order one and are cointegrated. **8** Moreover, using robust estimators in the presence of cross-sectional dependence in long-term economic activity, income inequality, trade openness, and innovative activity deteriorates sustainability, while renewable energy improves it in these countries.

1. What function do Sentences 1 and 2 perform?
2. What function do Sentences 3 and 4 perform?
3. Why is the present tense used in Sentence 5?
4. Can Sentence 6 be omitted? Why?
5. Does Sentence 7 work as a continuation of Sentence 6? And why?
6. Does Sentence 8 work as the conclusion of the abstract?

Task 7.11 Select two research papers in your field. Finish the tasks below.

1. Compare the titles of the articles in terms of length, structure, and information.

2. Analyze and compare the abstracts of the articles in terms of length, type, format, sub-headings (if applicable), and information elements.

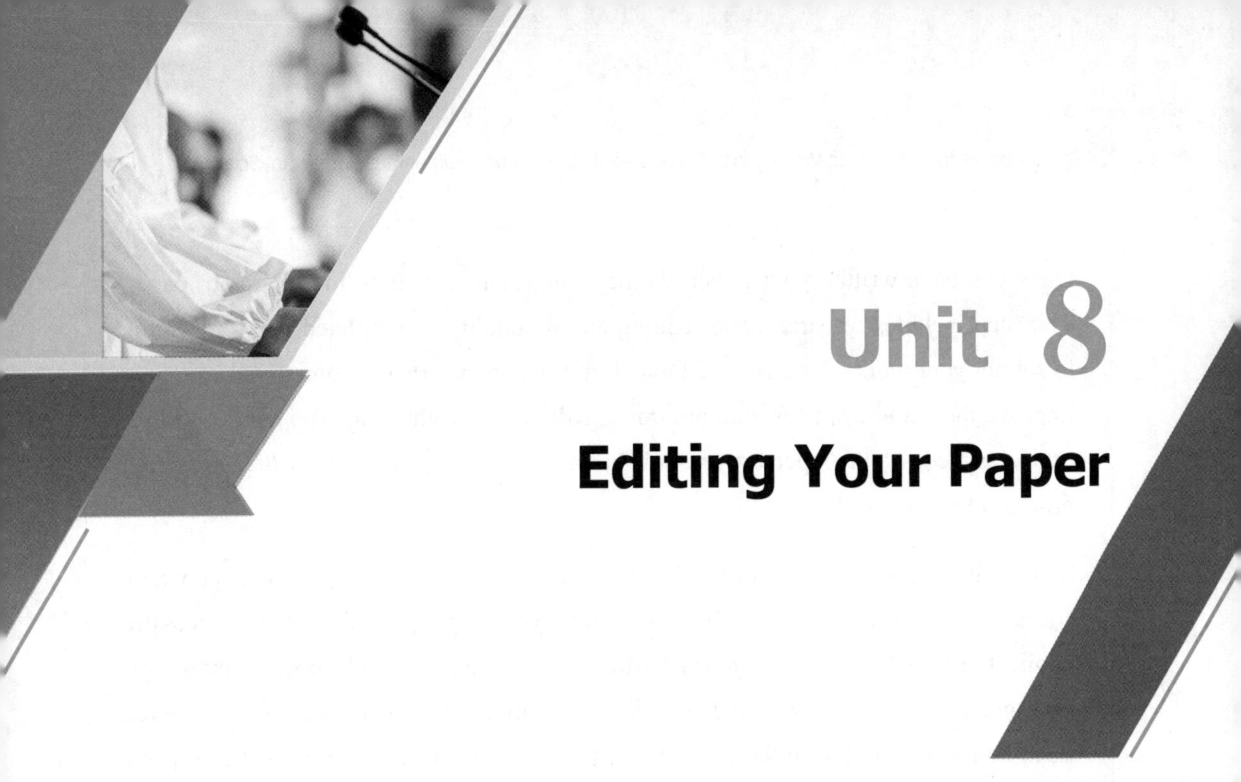

Unit 8

Editing Your Paper

Learning objectives

In this unit, you will

- grasp the fundamental concepts of manuscript editing; and
- formulate the fundamental approaches for manuscript editing.

Self-evaluation

At this stage, you may have finished writing your manuscript. Please consider the following questions.

- Are your ideas clearly and accurately expressed? Is there a logical structure to the ideas?
- Is the wording precise? Is the style formal enough?
- Does the text follow all formatting requirements?
- Are the grammar and sentence structures correct? How about the punctuation?

Once you have written your paper, the next important step is to improve your paper. This step includes research paper editing and it should be completed before deciding to submit your work. You may want to call in English experts or professional editors to improve the manuscript for you, but this is still not the right time. You can, and must, revise and edit your paper by yourself before you relinquish control of the writing process to other people.

Self-editing is the process of evaluating and making changes to your manuscript, which constitutes a crucial part in the course of getting your paper published. Most successful publications, including books, journal articles, news reports, novels, poems, and songs, have gone through several rounds of editing. No matter how significant your research is, to be successful during the peer review process, your manuscript must be properly edited.

Basically, there are three principles of self-editing, that is, conciseness, concreteness, and coherence (3C). In this unit, you will learn how to edit your paper by following these principles.

The 3C EDITING PRINCIPLES

Read the following passage extracted from the Introduction section of a research paper. Discuss with your partner whether it is easy to understand. If not, try to find out how to make it easier to follow by changing some sentences.

1 To enhance their global competitiveness, cities all over the world have been engaged in branding practices to construct their distinctive images (Kaika 2010), so that we have observed a lot of studies on city branding have proliferated in a variety of disciplines in recent years, such as tourism management (Francesconi 2011; Rabbiosi 2015, Dickinson et al. 2017), and urban policies (Löfgren 2014). **2** The majority of the studies are from the perspective of urban policy and planning, investigate the complexities of branded city images. **3** These studies have observed that similar to city identities in developed countries,

Chinese cities images are increasingly imbued with distinctive business characteristics—risk-taking, inventiveness, promotion and profit motivation (Ye and Björne 2018). **4** However, we still lack the understanding of how Chinese cities brand their identities, particularly using social media. **5** Addressing this need, the present study adopts a discourse analysis perspective and investigates the construction of city identities through systematic analysis of multi-semiotic resources in social media short videos.

Editing for conciseness

Owing to the nature of the topics being discussed, often technical and complex, academic writing should be clear, giving only necessary information, to the point, and describing things objectively. It will be helpful to keep the following in your mind while you are writing your paper:

- Using complex terms does not necessarily make you sound more intelligent. Instead, try to use simple, direct, but clear expressions. Cut everything that is not essential—this will let your key ideas stand out and be identified more easily.
- Being concise does not always mean using fewer words. It means using the least number of words that make the meaning 100% clear.
- The tense used for the verbs (particularly in the analytical process section) should be consistent.

Being concise is highly valued in writing, particularly writing for academic purposes. A word from the French writer and poet Antoine de Saint-Exupéry may help us appreciate the value of conciseness: "Perfection is achieved, not when there is nothing more to add, but when there is nothing left to take away."

The problem with conciseness may involve both content and language. It can appear within a single sentence or across a range of sentences. Unnecessary repetition of meaning, or redundancy, may confuse your readers and reduce their tendency to read the text in detail. The following sentence showcases this problem:

At this point of time, it now becomes necessary for us to consider alternative possibilities for the purposes of our goals.

This sentence is wordy and unclear, primarily because it is redundant. Some words could be deleted without affecting the meaning of the sentence. For instance, "at this point of time" and "now" mean the same thing, so the writer could save space by just using one of them. Then, the phrase "alternative possibilities" can be simplified into a single word "alternatives." In addition, "for the purposes of our goals" is unclear in meaning, and it also sounds awkward to use "purposes" and "goals" together. Finally, the writer could use the subject "we" to replace "for us" and a single verb "need" to take the place of "becomes necessary." Taking all the modifications together, the sentence will be revised as:

We now need to consider alternatives for achieving our goals.

By cleaning up redundancy, the sentence can be shortened from 21 words to just 10 words without changing its meaning. Compared with the original, the revised version is clearer and much easier to read.

Being concise means you give the information in as few words as possible. However, you may have frequently encountered long sentences in academic texts. It is true that long sentences are needed since research papers often deal with serious topics and convey ideas with high levels of sophistication. However, long sentences often contain multiple layers of ideas expressed in complicated structures, and could be hard to follow and even cause confusion. Although you may want to use long sentences to increase the professionalism and formality of your text, you still need to give priority to clarity. Sometimes, you may need to break a long sentence into shorter ones.

Task 8.1 Revise the following text to make it more concise.

As noted by Pan (2021: 9), by offering an effective tool for "evoking emotions of grief, anguish and hatred to mobilize resistance and attest to a moment of national awakening," such direct textual description of the devastation caused by the western attacks can foster a sense of popular nationalism among the audience (legitimising their sense of duty to the past), and is consciously harnessed by the Party-state as a way to counter elements of western political values.

Editing for concreteness

Concreteness in an academic context largely means being specific, definite, and vivid rather than general and repetitive, i.e., providing specific details for what you are writing about. The details can be facts, examples, statistics, or citations. Sometimes, novice writers err in their papers by not providing such specific details, which makes their writing general and dry. Even if a paragraph is well structured, vague wording can confuse readers and make them doubt your credibility.

Vague modifiers: A modifier qualifies the meaning of a noun or verb. Words such as *reasonably*, *extremely*, *rather*, *difficult*, *ideal*, *very*, *really*, and *well* do not qualify the meaning specifically.

Vague *To guarantee the validity of the coding, the two authors co-coded the videos independently and the results were reasonably compared.*

Precise *Both intra-coder and inter-coder reliability tests were applied to ensure the accuracy of the coding. The film was coded twice by one of the authors at an interval of three months; the two authors co-pilot-coded 10% of the film independently and the inter-coder reliability was above 80%. All inconsistencies and ambiguous cases were discussed and resolved by the two authors.*

Vague references to research method: A vague mention of the research method does not well inform the readers of how you conduct the study step-by-step, which may hurt the credibility of your work.

Vague *Based on the close reading, we generated the open codes by using the protagonists' lines and their visual depictions (i.e. vivo coding).*

Precise *Following the grounded theory approach, the coding involves two steps: (1) The open coding of attributes in the film, and (2) the classification of attributes into categories (i.e. axial coding). An instance is recorded each time when a character's attribute is represented in language (by explicitly using evaluative lexis or implicitly recounting facts) or visual images (by the depiction of characters and background settings).*

Vague references in analysis: If you want to be certain that a reader will understand your argument, follow it up by providing a concrete example of what you are talking about. In the example text, it needs to add more robust explanations by adding more examples to make the interpretation more convincing.

Vague *The ubiquitous female pejoration led to a sense of blunt sexism on social media.*

Precise *The ubiquitous female pejoration in the online community, for example, has led to blunt sexism and perpetuates a gender order the quintessence of which is a relationship of subservience and domination.*

Vague references in discussion: The finding text sometimes can be interspersed with what seem to be generalized statements rather than discussing what the findings illustrate. At depth, it does not provide what needs to engage the reader in interpreting the findings. Sometimes, the discussions may tend to be predominantly attributed to the seemingly already established literature; one may wonder what the paper's true contribution is?

Vague *The pattern of attributes is finally explained in relation to the social context of the* wanghong *economy and the affordance of TikTok.*

Precise *In the digital age where the Internet, and social media in particular, has transformed how we communicate and consume, new forms of institutional discourse are emerging rapidly. Multi-disciplinary and contextualized theoretical accounts are needed to understand their new meanings and the complex semiotic resources for realizing the meanings. This study is a modest step towards such an understanding and it is hoped that it can inspire further semiotic studies on various forms of digitalized institutional discourse in new context.*

Overall, you need to provide specific details to support your claim and avoid using vague expressions in your writing. When you edit your draft, keep asking yourself:

- Have I included specific details to enable a clear understanding in readers?
- Have I provided sound evidence for the credibility of my work?
- Have I had any vague wording which may result in confusion?
- Can I substitute an exact statement or a figure for a general word to make your message more concrete and convincing?

Task 8.2 Compare the two versions of the same text below. Underline the parts that are different. Think carefully about which version you like better and why you like it.

Version 1

The city under investigation is Xi'an, which was chosen for its successful rebranding of its city image by deploying rich symbolic resources and leveraging social media. As a traditional industrial mid-western city, Xi'an hit a bottleneck in the urban transformation process and has been seeking new urban imaginaries in response to intensified inter-urban competition in the new millennium. In 2018, Xi'an Tourism Bureau began to modernize its city image and established a full-scale cooperation plan with TikTok, a leading social media platform for creating and sharing short videos in China. The collaboration, which resulted in the effective exploitation of symbolic resources to transform its city image, made Xi'an "the most popular *wanghong* city" in China, surpassing tier-one cities like Beijing and Shanghai (Liu 2019). According to the *White paper on city images and short videos on social media*, up till December 2018, there were over 1.9 million TikTok videos related to the term "Xi'an" in hashtags, and the overall video views reached nearly 9 billion (Chinese Internet Data Centre 2018). The benefit of creating a *wanghong* image in the virtual world immediately manifested itself, with a 50% increase in tourist revenue in 2018 (Li and Jiang 2018).

Version 2

The city they're talking about, Xi'an, got all fancy and stuff by using symbols and doing the social media dance. Xi'an used to be a plain old city in the middle of China but then it got stuck in a makeover mess and had to find new ways to look cool because all the other cities were flexing hard. So, in 2018, the Xi'an Tourism Bureau thought, "Let's get trendy!" and hooked up with TikTok, the video app everyone's crazy about in China. This tag team made Xi'an the coolest city for Internet celebs, beating out big shots like Beijing and Shanghai (Liu

2019). Some official reports said that by the end of 2018, there were like a bazillion TikTok videos with "Xi'an" in the title, racking up billions of views (Chinese Internet Data Centre 2018). And guess what? This online fame brought in cash, with tourist money jumping up by 50% in 2018 (Li and Jiang 2018).

Editing for coherence

Generally, the extent to which writing "flows" is referred to as coherence. Coherence is the result of tying information together so that connections you have made in your own mind are apparent to the reader. Research papers contain complex technical information that often needs to be expressed in long and complicated sentences, which may increase the difficulty for readers to follow the text. Common solutions to this problem include relating each sentence in the body of a paragraph to the topic sentence and linking each sentence to a previous sentence. In doing so, you make your writing flow smoothly and lead your readers to access and process the intended information as the text unfolds. There are usually four ways to achieve text coherence.

1. By using connectors that help establish the sequence of events

Conjunctions or conjunctive adverbs such as *therefore*, *accordingly*, *in this regard*, *by doing so*, and *in this way*, are widely used to make sentences logically and grammatically more coherent.

2. By using recycled words

Repeating the keywords may help linking ideas and reminding readers of the core information or central idea.

3. By following the old-to-new pattern

Start a sentence with information that readers already know, such as knowledge shared by your readers or the elements already mentioned in the previous text. Then, new information is introduced in the rest of the sentence. It will be easier for readers to digest information if they move from what is known (old) to what is to know (new). For example:

In order to enhance their global competitiveness, cities all over the world have been engaged in branding practices to construct their distinctive images (Kaika 2010). Accordingly, studies on city branding have proliferated in a variety of disciplines in recent years, such as tourism management (Francesconi 2011; Rabbiosi 2015, Dickinson et al. 2017).

4. By moving from general concepts to increasingly more specific concepts

Another important aspect regarding text readability is the logical progression within the text. Conventionally, general concepts are provided first, followed by more specific concepts. A view of the big picture will help readers understand specific details. The following example showcases the general-to-specific pattern of text organization:

Like product branding, the "cultural turn" in urban studies has directed urban sociologists' attention to the concept of "symbolic economy" (Zukin 1996), which points to the material commodification of cultural meanings that are attached to specific places. An important concept derived from social imaginary and particularly useful for our study is "urban imaginary" (Zukin et al. 1998; Greenberg 2000, 2003). According to this concept, a city is essentially a space of cultural production and consumption, and ideal city brands are realized through the deliberate manipulation of the symbolic and emotional values.

Task 8.3 Read the following paragraph. Underline words and phrases that contribute to the coherence of the text.

Adopting a critical multimodal discourse analysis method, we develop a semiotic framework to model Xi'an's digital image found in TikTok video clips as a set of evaluative attributes. Thus, our specific research questions include: (1) what are the distinctive urban imaginaries constructed in Xi'an's official TikTok videos, (2) how are the urban imaginaries realized through the use of linguistic and visual resources, and (3) what do the features of urban imaginaries reveal about urban policies in China. In what follows, we will first introduce our theoretical basis of symbolic economy. We will then describe our data and analytical framework, which will then be followed by an analysis of the urban imaginaries and their multimodal realization. Finally, the results will be discussed in relation to the current urban policies.

Check your understanding

Task 8.4 Find a journal article related to your study. Read several paragraphs carefully and analyze the text in terms of the 3C principles.

Unit task

Peer Review

After you finish your draft, you may need to have someone else read your paper and give you feedback. This practice is beneficial as the readers may identify problems that you do not realize. In this task, you will work with your partner and review the Introduction section for each other. Do the following to finish this task.

Step 1: Check each paragraph.

- A topic sentence is upfront showing the focus of the paragraph.
- All the rest sentences are relevant to the focus.
- Ideas are arranged in the general-to-specific pattern.
- Specific information is provided so that readers may have a good understanding and not be confused.
- Sentences are concise and easy to understand.
- Sentences are well connected and information flows naturally and smoothly.

Step 2: Leave comments.

- Make comments on your peer's work by using the "Comment" function in MS Word. The comments can be appreciation or agreement on what your peer has done well, and suggestion or recommendation for what could be improved.
- Provide revision suggestions by using the "Track change" function in MS Word. Make sure that your peer can access both the original text and your revisions.

Step 3: Revise the peer-reviewed draft.

- Revise your draft based on your peer's feedback.
- Use a grammar-checking tool, such as Grammarly to avoid grammatical or logical errors.

Unit 9

Submitting Your Paper

Learning objectives

In this unit you will

- understand the selection criteria of the target journal;
- learn how to write a cover letter; and
- learn how to respond to reviewers' comments.

Self-evaluation

At this stage, you may have completed your manuscript. It is time to think about the following questions concerning how to submit your manuscript.

- What are your target journal's requirements for manuscripts? How to format your research paper?
- How to write a cover letter when submitting your manuscript?
- What are the general principles of responding to reviewers' comments?

Submitting your manuscript to a journal for publication is a long and complicated process that brings you great anxiety and stress. Throughout this process, you may have to negotiate with journal editors and referees back and forth through emails and work hard on revising and editing your manuscript based on their suggestions. Whether your manuscript could be accepted for publication is determined by journal editors and referees using a defined set of selection criteria. Meeting the selection criteria of the journal is thus key to publication success. In this unit, you will learn the selection criteria of your chosen journal, how to write a cover letter, and how to respond to reviewers' comments on your manuscript. All the information will help you develop your publishing strategy and navigate the publishing process, leading to publication success.

SELECTION CRITERIA OF THE TARGET JOURNAL

Select a journal in your field, to which you may submit your paper in the future. Try to find out specific requirements your manuscript should meet before your submission. Take the following three steps to complete this task.

Step 1: Go to the journal website to find author guidelines.

Step 2: Note down the requirements you feel important or you did not know before reading the guidelines.

Step 3: Exchange what you find with your partners.

Target journal's requirements for manuscripts

A well-organized manuscript will be easier for reviewers to understand and will help them to appreciate its impact. Therefore, the basic selection criterion of a journal is related to the writing style of manuscripts. Most journals provide detailed instructions on how to draft and format papers. Such instructions can be found on the homepage of a journal, often titled as "Author Guidelines," "Guide for Authors" or "Instructions for Authors." Please always review the style guide of your target journal before submitting your article.

Task 9.1 The table below is from the formatting guidelines for authors of a journal. Suppose you would submit your manuscript to one of this journal. Discuss with your partners what you should do before your submission.

Elements	Requirements
Title	Informative, accurate, and concise description of the manuscript summarizing your main results
Abstract	• Problem and objectives • Methodology used • Findings and conclusions • Research's effects on broader scientific issues ***Note:*** While each journal may have its own requirements for abstract length, the norm is typically 200–250 words.
Keywords	• Three to five keywords that collectively describe your research
Introduction	• Problem to be addressed • Background and literature review • Research purpose and method
Main Body	• Problem, assumptions, and limitations • Theoretical foundations, framework, and methodologies • Analysis, findings or results of the study
Discussion	• Discussion of results and how findings reflect upon the larger context of the research field • Comparison of results with other related work • Significance of results

Continued

Elements	Requirements
Conclusion	• Summary of specific conclusions • Limitations and relevant issues for future consideration
References	• References numbered in order of appearance • Following the journal's referencing style guidance *Note:* Reference management programs can be useful (e.g. EndNote & Mendely).
Appendices	• Supplementary material for completeness which could detract from the logical presentation of the work • Material that could be valuable to specialists or those wishing to reproduce your results
Acknowledgements	• Technical assistance and useful comments (e.g. from colleagues, advisors, reviewers etc.) • Financial support and other relevant disclosures

Task 9.2 Fill in the table below according to the requirements of your target journal. Discuss with your partners any requirements that you think important, confusing, or unexpected. You may discuss questions such as:

- **How should I balance "be concise" and "be informative" in writing my title?**
- **I haven't seen many titles starting with "the." Why does the title not begin with an article?**
- **The words "first," "new," or "novel" seem to claim the novelty of a study. Why should they not be placed at the beginning of the title?**

Elements	Requirements
Title Page	Clearly state author names, department, university and country; provide full correspondence details and short biographical notes on all contributors

Continued

Elements	Requirements
Title	Concise but informative. Don't begin with an article, a preposition, or the words "first," "new," or "novel"
Abstract	
Keywords	
Subject classification codes	
Section headings	
Sections (what sections are required?)	
References and notes	
Funding	
Acknowledgement	
Disclosures	
Figures	
Tables	
Supplementary materials	
Video files	
Others	

Reviewers' evaluation

All the submitted research articles must undergo a peer-review process before they are selected for publication in journals. Peer review has been defined as a process of subjecting an author's scholarly work, research or ideas to the scrutiny of others who are experts in the same field. It functions to encourage authors to meet the accepted high standards of their discipline and to control the dissemination of research data to ensure that unwarranted claims, unacceptable interpretations or personal views are not published without prior expert review. Reviewers' comments not only help a senior editor decide on whether to accept or reject the paper, but also work as a useful source of feedback, helping writers to improve their paper before publication.

Reviewers' proposal to the senior editor may be to accept the paper as presented, to make changes as recommended by reviewers and repeat the review process, or to reject the paper.

A journal usually takes several months to complete the review process, which typically involves

- reading the article and deciding whether to send it for review;
- acquiring sufficient reviewers and receiving all feedback; and
- assessing the reviews and rendering a decision on the paper.

Task 9.3 The following texts are the proposals of some reviewers to the senior editor of a journal. Read them carefully, underline the sentences indicating reviewers' proposals and fill in the blank below each text.

Text 1

In the revised manuscript, the authors have addressed all my major concern. Other comments are also well-addressed. I think the manuscript in its present form is suitable for publication in *Environmental Communication.*

Reviewer's proposal: ____________

Text 2

I read this article with great interest. The authors have meticulous explained the relevant concepts and findings in the literature review and results section. Most parts are also well-structured. I have a few suggestions on how to improve the manuscript.

(1) As critical discourse analysis is one of the focuses in this study, there needs to be more discussion of it in the literature review.

(2) Some sentences are somewhat unclear and not idiomatic, such as "demonstrating their sexualized bodies" (p.3).

(3) More details about coding are needed.

Reviewer's proposal: ___________

Text 3

The manuscript presents an interesting study of eco-documentaries in China and its findings have the potential to complement ongoing scholarly discussion among *Environmental Communication*'s authors and readers. This said, I found that the manuscript requires major revisions in terms of how its theoretical and empirical concerns are connected with existing environmental communication scholarship in the Western context. Specifically:

(1) The manuscript begins by noting that "studies on environmental branding of a nation have proliferated in a variety of disciplines in recent years." Yet, details of these theoretical perspectives have not been sufficiently elaborated in Section 2. The current manuscript needs to be in conversation with such articles so that its theoretical implications would make sense for EC's international readership.

(2) More on the Methods section. Appraisal theory and Systemic Functional Grammar need more explanation for readers from non-linguistic background. Moreover, you need to provide details as to the number of coders and statistics in inter-coder reliability. This is standard for content analyses.

(3) Echoing my previous comment on Section 2's need for more theoretical engagement with existing environmental communication literature, the final section does not effectively elaborate on the study's theoretical implications for the advancement of EC research. How would the semiotic representations in Chinese eco-documentaries offer valuable lessons for fellow environmental activists and practitioners in other countries? I recommend providing more reflections on questions like this.

Reviewer's proposal: ___________

Text 4

I appreciate that the authors did spend some effort addressing the concerns about the previous manuscript. However, the overall manuscript is unsatisfactory and unsuitable for publication. Specifically:

(1) Overall, the manuscript needs English-language editing. At times the wording makes it hard to follow the main points in the narrative. Be sure to write in directly, concise, clear terminology and use active voice throughout.

(2) The first section of the article appears to be the introduction, which should more clearly declare the problems of the researcher(s). At the beginning of the article, a large amount of space is devoted to discussing the event. However, it's hard to figure out what kind of social issues are reflected in this background information. Why do we need to understand the message Modi delivered in his speech? And why do we need to know the rhetorical devices and the logical organization he utilized to deliver his message? What is the knowledge gap?

(3) Details of these theoretical perspectives have not been sufficiently elaborated in Section 2. It is difficult to understand how this literature relates to the study design, data analysis, and research discussion that follows. The author(s) may need to increase the structure of this section, clarify the core theoretical perspectives and key concepts of the article through literature review more purposefully, and apply them to subsequent analysis.

(4) The organization of the paper is chaotic. The next two sections were supposed to be part of the Methods section.

(5) The whole analysis part is a bit too subjective. I think the author(s) need to consult relevant scholarly sources to make their analysis more evidence-based and objective.

(6) The conclusion lacks any substantive material; it reads more like a "summary" rather than a "conclusion." What generally this section needs is to deliver what the paper intended to do at the outset. This is the second glaring deficiency to which I would like to direct the authors' attention—one which hurt the paper's contribution. Perhaps, a lack of a theoretical underpinning at the beginning manifests itself here as the authors sufficed to summarize the paper in a rush.

Reviewer's proposal: ____________

COVER LETTER

Below is a cover letter. Read it and specify what each paragraph is about.

Dear Editor,

1 It is my pleasure to submit an original article "History, modernity, and city branding in China: a multimodal critical discourse analysis of Xi'an's promotional videos on social media" to *Social Semiotics* as the corresponding author.

2 To enhance their global competitiveness, cities all over the world have been engaged in branding practices to construct their distinctive images. Against this backdrop, this study aims to investigate how Xi'an, a second-tier developing city in China, constructs its digitalized urban imaginary using the popular social media platform TikTok. A semiotic framework is developed to model Xi'an's urban imaginary as evaluative attributes and to elucidate how they are constructed through linguistic and visual resources in short videos on TikTok. The analysis of 294 videos shows that Xi'an highlights its dual identity as a modern metropolis and a historical city. The modern metropolis image is characterized by the personification of Xi'an as a stylish, young, popular, and international microcelebrity; the historical city image is constructed by recreating the Great Tang Dynasty and revitalizing local folk art. The characteristics of city branding discourse reflect China's *wanghong* economy, urban policies, and the affordances of social media.

3 We confirm that the work described has not been published previously, that it is not under consideration for publication elsewhere, and that, if accepted, it will not be published elsewhere without the written consent of the copyright-holder.

4 Thank you for your consideration. We are looking forward to hearing from you with the updated information soon.

Sincerely Yours,
Dr. Peng Wang

Paragraph	What it is about
1	
2	
3	
4	

Contents of a cover letter

Writing a cover letter is an essential part of the journal submission process. A strong cover letter can impact an editor's decision to consider your research paper further and ultimately determine whether to publish it in their journal. A cover letter usually includes the following essential information element.

IE1: Specification of your manuscript

You should first provide basic information about your manuscript: the title and the type of your paper (e.g. original article, review article, case report, communication etc.).

IE2: Brief summary of the paper to convince the editor

You may briefly describe the rationale and background of your study and the reasons why your paper deserves being published in the journal. You need to convince the editor that your findings are of significance, theoretically and/or practically.

IE3: Declaration of conflicts of interest

You may provide information about conflicts of interest that involves any author. You may also include the sources of outside support for research, such as funding and facilities. If there was no conflict of interest, just write: "No potential conflict of interest relevant to this article was reported."

IE4: Information about dual submission or prior publication

All journals set policies regarding dual submission or resubmission of previously published papers. You may declare in the cover letter that your paper has not been published elsewhere or submitted to other journals for consideration of publication.

IE5: Contact information

Include detailed information of the corresponding author (affiliation, mail address, e-mail address, and telephone number).

Task 9.4 Select an article of your interest from a journal you follow. Pretend that you were the author and would submit the article to the journal. Scan the article to get necessary information and write a cover letter.

RESPONSE LETTER

Below is a response letter to the editor's feedback. Read it and specify what each paragraph is about.

Dear Dr. Rachel Griffin,

1 We have already finished our revision and re-submitted it (Manuscript ID: XXX) on the platform. We really appreciate the two reviewers' detailed and insightful comments, which greatly enhanced the clarity and coherence of our work.

2 The revised manuscript via Track Changes, together with a brief response to the editor, is attached below. We have carefully modified the manuscript in order to clarify the comments raised by the two reviewers. Please find below our detailed answer to the referees. The major revisions in the revised manuscript are marked in red. We hope that, after these modifications and explanations, the manuscript will be suitable for publication in Journal of XXX.

3 Thank you very much for your efforts. We are SO looking forward to seeing the successful publication of our manuscript in CDMC.

Best regards,

XXX (on behalf of all authors)

Paragraph	What it is about
1	
2	
3	

Response to the editor's feedback

After weeks of waiting, you will receive the editor's decision letter. There are several possible decisions made by an editor.

- **Accept:** The paper is accepted for publication without any further changes required from the authors.
- **Minor revision:** The paper is accepted for publication in principle once the authors have made some revisions in response to the referees' reports.
- **Major revision:** A final decision on publication is deferred, pending the authors' response to the referees' comments.
- **Reject and resubmit:** The paper is rejected because the referees have raised considerable technical objections and/or the authors' claim has not been adequately established. Under these circumstances, the editor's letter will state explicitly whether or not a resubmitted version would be considered.
- **Reject:** The paper is rejected with no offer to reconsider a resubmitted version.

The editor's decision letter usually encloses reviewers' comments, to which you are advised to respond one by one.

Task 9.5 Below is an editor's e-mail to a writer. Suppose you were the writer and write a letter responding to the editor's e-mail.

Dear Professor XXX,

Thank you for submitting your manuscript entitled "XXX" for publication in journal of XXX. The review process has now been completed. The reviewer comments are included at the bottom of this letter.

The reviewer(s) would like to see some revisions made to your manuscript before reconsidering it for possible publication. I agree with their assessment, and invite you to respond to the reviewer(s)' comments and revise your manuscript.

In accordance with our format-free submission policy, an editable version of the article must be supplied at the revision stage. Please submit your revised manuscript files in an editable file format.

To submit a revision, go to the website XXX. If you decide to revise the work, please submit a list of changes or a rebuttal against each point that is being raised when you submit the revised manuscript.

If you have any questions or technical issues, please contact the journal's editorial office at XXX@journals.com.

Once again, thank you for submitting your manuscript to journal XXX and I look forward to receiving your revision.

Sincerely,

Dr. XXX

Editor, XXX

Response to reviewers' comments

A proper response to reviewers' comments on submitted articles is essential to publication. Below is the essential information that should be included in your response letter and be taken into account when you write the letter.

- Title of the manuscript.
- A brief "thank you" note addressed to the editor and reviewers stating your gratitude for the review.
- Write responses separately to the comments from different reviewers.
- Format the letter in a way that your responses are distinguished from the reviewers' comments.
- Be sure to answer each and every comment made by the reviewers. This is often called "point-to-point responses to comments."

There are many ways to deal with reviewers' comments, and you will develop your own strategies. Here we outline an approach used by many experienced authors.

- Make all the changes as required or suggested and note each change in your letter of responses.
- Avoid taking a strong or argumentative tone if you happen to disagree on any comments. Instead, state that the reviewer has raised a good point, try to argue in a more positive tone why you do not agree, and provide as many facts as possible to support your argument.
- If you do not agree with a reviewer when he or she recommends, for example, another method to be used, you need state clearly why you think the original one is better and provide references if possible.

Task 9.6 Below are some reviewers' comments. According to the given instruction, respond to the comments properly.

1. Reviewer's comment: The data seem to be more about print media (online news articles). Have the authors considered non-traditional media such as social media?

   Response (not agree with the reviewer): ____________________________________

2. Reviewer's comment: The expression of the keyword "Chinese-language news media" is inconsistent with the "Chinese English-language news media" mentioned in the title and the body of paper, which appears confusing to the reader.

   Response (agree with the reviewer): ______________________________

   ____

3. Reviewer's comment: I think that the findings need to be presented as answers to the research questions. It is important to mention the research questions when presenting the findings. The findings should be organized around the research questions.

   Response (agree with the reviewer): ______________________________

   ____

Check your understanding

Task 9.7 Please exchange ideas with your partner(s) on your manuscript, and try to reply to the comments from your partner(s). You may use expressions such as:

- **We thank the reviewer for pointing this out. We have revised...**
- **We have removed...**
- **We agree and have updated...**
- **We have fixed the error...**
- **This observation is correct...we have changed...**

Task 9.8 Find a person (your peers or professors) who has succeeded in publishing an article in an English journal, ask for all the documents (cover letters, editor's feedback, reviewers' comments, and response to reviewer's comments) involved in the process, and share them with your classmates. If you yourself have such an experience, just share it with your classmates.

Writing a Cover Letter

After you finish editing your manuscript, you are ready to submit it to the target journal. In this task you will write a cover letter according to what's offered in this unit. You may also use the following template to finish the task:

Salutation	Dear Dr./Mr./Ms. [Editor's last name]:
Paragraph 1: Purpose and subject	I am writing to submit our manuscript entitled [Title] for consideration as a [Journal Name] [Article Type]. [One to two sentence "pitch" that summarizes the study design, where applicable, your research question, your major findings, and the conclusion.]
Paragraph 2: Appropriateness to the journal	Given that [context that prompted your research], we believe that the findings presented in our paper will appeal to the [Reader Profile] who subscribe to [Journal Name]. Our findings will allow your readers to [identify the aspects of the journal's Aim and Scope that align with your paper].
Paragraph 3: Additional statements often required	Each of the authors confirms that this manuscript has not been previously published and is not currently under consideration by any other journal. Additionally, to the best of our knowledge, the named authors have no conflict of interest, financial or otherwise.
End	Sincerely, [Your Name] Corresponding Author [Institution/Affiliation Name] [Institution Address] [E-mail address]

Unit 10

Presenting at Conferences

Learning objectives

In this unit, you will

- understand the features of the conference communication;
- develop the strategies for making poster and slide presentations at conferences; and
- practice the skills and techniques of delivering oral conference presentations.

Self-evaluation

How could you gain quick access to an academic conference in your field of interest? Scan the QR code to access the poster for the call for papers at an international conference. Read it and answer the following questions.

- When and where would this conference be held?
- When could the participants register?
- How could participant submit their conference papers?
- What topics would be discussed in the conference (list at least 3 topics)?

Academic conferences, whether national or international, online or in-person, play a crucial role in facilitating academic exchanges. Researchers attend these conferences to showcase their research and engage with their peers. To participate in an international conference, it is essential to prepare an English research article, communicate with the conference organizer, create PowerPoint slides and posters that meet academic standards, give a poster or oral presentation at the conference, and engage with fellow participants. As a presenter, it is important to present your work clearly and respond effectively to audience questions. As an audience, providing constructive feedback on presentations and engaging in discussions are also expected. Developing relevant skills and techniques is essential for fulfilling these roles effectively.

This unit is designed to offer activities and tasks that will help you acquire the necessary skills and techniques for presenting your research visually and verbally at an international academic conference.

VISUAL DELIVERY

Scan the QR code. Read the poster about research on teaching statistics in psychology and answer the following questions.

1. What are the major components of the poster?

2. What type of information is visually the most salient?

3. What do you think about the design of the poster, such as the use of colors and the layout?

4. How does the poster attract and engage the viewer?

Designing a poster

Posters are an essential part of scientific communication. They serve as advertisements for both your research work and yourself. The traditional format is a printed poster displayed on a poster board in an exhibit hall. However, many conferences now use electronic posters (e-posters). Designing an effective poster can be challenging due to the limited space, typically around 100cm×200cm.

The process of designing a poster involves planning, selecting, drafting, creating, and reviewing. It should include a title followed by sections on objectives, methods, results, and discussion/conclusion. A successful poster effectively presents and summarizes the main points of your research, with additional details provided by the presenter during the presentation. On the other hand, an enlarged copy of your manuscript is the least effective option. It is recommended to use tables and figures to enhance the presentation of data and make it more engaging.

Your poster should include

- the paper title and all authors at the top;
- a brief introduction, goals, experimental details, conclusions, and references, presented in a logical and clear sequence; and
- tables, graphs and pictures, with an explanation for each of them.

While designing the poster, it is advisable to use templates (scan the QR code to read a sample) provided by conference organizers or universities, but also to be creative in order to create a visually appealing design. Try to convey your messages using minimal text and strong imagery. Consider factors such as layout, color, background images, text and graphics placement, format, readability, and scannability. It is important to check the dimensions of the poster boards in advance. Once the entire poster is assembled, it is recommended to print out a draft copy to see how it looks in reality, as it may differ from the computer-generated version. Gathering feedback from as many people as possible is also beneficial.

Task 10.1 Scan the QR code to access a sample poster. Analyze the poster design and give your comments on it in the table below.

Poster design	Comments and suggestions
Format	
Text	
Graphics	
Colors	
Readability	

Task 10.2 Select a journal paper of your interest and design a conference poster for the work. Bring your poster to class and share with your classmates.

Producing presentation slides

It is nearly impossible to envision an academic presentation being given without the support of a computer-based slide show. A computer slide show enables a presenter to pair visual aids with verbal explanations. A picture is worth a thousand words, and the retention of information tends to be higher when spoken words are accompanied by visuals. Therefore, the combination of visual illustrations and explanatory text is the most effective way to convey information accurately and efficiently. Well-designed slides that are suitable, precise, legible, understandable, well-crafted, engaging, and memorable are essential tools for conference presenters.

When creating slides for a scientific presentation, emphasis should be placed on clarity and impact. Every element on the slides should contribute to conveying the intended message; any irrelevant content should be eliminated. Here are some tips to keep in mind when designing your slides:

- Understand your audience.
- Ensure your slides are easy to read.
- Utilize high-quality images and graphics.
- Organize your presentation logically.
- Present one idea or point per slide.
- Keep transitions simple.
- Establish a consistent color scheme.
- Choose appropriate fonts.
- Incorporate vector icons where needed.
- Include ample "white" space.

While visual elements are crucial, text is often necessary on slides, especially when outlining study objectives, providing definitions, summarizing or concluding, and displaying quotations. The following tips can assist you in creating text slides:

- Use standard written English.
- Adhere to proper grammar rules.
- Select the right words.
- Maintain consistency in tense, tone, and style.
- Proofread for spelling errors.
- Be clear, concise, and accurate.

Task 10.3 According to Guy Kawasaki's 10/20/30 Rule, an effective presentation should be ten slides long, last for 20 minutes, and have a font size of 30 points. Does this rule work in your research field? Why or why not?

Task 10.4 **Scan the QR code to access the PPT slides of a presenter at the Workshop on Environmental Governance in Asia. Read the slides and answer the questions below.**

1. What are the major components of this slide presentation?

2. What do you think of the lexical correctness of the texts on the slides?

3. Which part of the presentation is heavily loaded with graphics? Why?

4. Are the slides easy to read? Why or why not?

ORAL DELIVERY

Below is the welcoming speech made by BAN Ki-moon, Chairman of the Boao Forum for Asia, which was delivered at the Annual Conference 2024. Read it and fill in the table with the paragraphs of the speech that perform the given functions.

1 Welcome to the Annual Conference 2024 of the Boao Forum for Asia in the sunny, mild and promising springtime of the Year of the Dragon.

2 Welcome back to this annual gathering of the Boao Family to renew friendship, reconnect, review, debate, plan and act. More than 2000 government, business and intellectual leaders from Asia and around the world are gathered in this tiny island for a three-day intensive brainstorming on the most pressing issues of the day. People lay high hopes on us. We must not let people down.

3 Looking back, there is much to take comfort from and pride in. We have beaten the worst-case scenario, and emerged from a successive barrage of shocks and crises with stronger-than-expected resilience. The world is freed from the grip and fear of the pandemic. Inflation is less threatening as we thought it would be, and falling fast in most countries. Global economy may well escape a hard landing. A modest but steady rise is in sight.

4 Looking ahead, however, it's no all roses or plain sailing. Challenges abound and weigh heavy. Black swans, grey rhinos, white elephants or black jellyfish may pop up anytime on the way. Quite often, in a highly connected and globalized world, these challenges and risks are not confined to one country or region. They cross borders, spill over, and force us into one same boat in the unchartered waters.

5 The climate crisis, for instance. None of our economic or geopolitical headaches has disrupted the pace of global warming. It just keeps going up. The past year of 2023 was just confirmed as the hottest year on record. That record, however, may well be broken again as January 2024 was just found to be the hottest month on record.

6 The world is acting, but not fast and enough. Last December, the COP28 vowed to transition away from fossil fuels, triple renewables and increase climate finance for the most vulnerable. A historic agreement to keep the 1.5°C goal alive, but just the beginning, not the finish line. Global climate action must follow up and match the words with rapid, deep and sustained reductions in the global greenhouse gas emissions of 43 percent by 2030 and 60 percent by 2035 relative to the 2019 level, and net-zero carbon emissions by 2050.

7 Global action must be based on trust and solidarity, which unfortunately is quickly missing in the past few years. On the rise are intensifying rivalry, confrontation and conflicts. The Ukraine conflict and the Israel Palestinian conflict threaten to rip the world further apart. Countries are regrouping and realigning along ideological or geopolitical lines. We're forced into one same boat, but not everyone rows in the same direction. Sustained peace and sustainable development, two lofty goals underpinning the United Nations, are still far away.

8 Addressing such increasing complexities and uncertainties takes vision, courage and innovativeness which, however, must build on common sense and proven wisdom. We must go back to the proven values and principles enshrined in the UN Charter. We must rekindle the proven institutions with the UN at the core. We must revitalize the multilateral approach that has helped safeguard peace and promote prosperity. In the face of common challenges, we must row in the same direction as a team.

9 As former Secretary General of the UN and Chairman of the Boao Forum for Asia, I have deep faith in the power of countries and people working together. I call upon you, leaders and fellow delegates of the BFA Annual Conference 2024, to rise to our common challenges, take up our shared responsibilities and work together towards our shared future of peace and development. Your stay in Boao is only four days, but enough to make a big difference in Asia and the world.

10 I wish the Annual Conference 2024 a shining success as always.

Note: This welcome speech is taken from *2024 Boao Forum for Asia Annual Conference Journal* on the website of https://www.boaoforum.org/ac2024/html/list_2_602_1.html.

Function	Paragraphs
To introduce the background, theme and aim of the speech	
To illustrate the problems and challenges	
To explain the measures and approaches	
To express thanks and best wishes	

Making poster presentation

When you arrive at the meeting, you will need to hang up (or upload) your poster. It is important to check the specific conference website for your poster numbers, location, and setup times. Some meetings provide pushpins for hanging up posters, but this is not always the case, so it is best to check in advance. Generally, the presenter of the poster should stand by the poster, at least during the mandatory time slots, and preferably more frequently. Failure to comply with these rules may result in disqualification from future meetings. The most popular times for people to view posters are during lunch breaks and poster sessions. To prepare for any questions from the audience, it may be helpful to show the poster to colleagues or present it at a lab or research meeting at your own institution before the scientific meeting. This can generate valuable comments. Special judges may be assigned to score posters for awards. Their evaluation is usually based on the overall presentation, including the abstract, the poster itself, and the attentiveness and presence of the presenting author.

Additionally, it is important to follow the established dress code for the meeting. If you are unsure about the dress code, you can ask someone who has attended the meeting before or contact the organizers directly. It is always better to be slightly overdressed, as you are representing yourself, your research, and your institution.

Finally, you may consider adding a handout to your poster that attendees can take with them. This handout could include the abstract, key points, and the contact information for the corresponding authors. Some meetings may make the abstracts available to registered attendees online, on a flash drive, or in a printed program book.

Task 10.5 Read the poster guide of the ISPOR 2023 conference for advancing the science and practice of health economics and outcomes research around the world. Make your adjudgment and decide whether each of the following statements is true or false.

Poster Guide

Poster Discussions

Posters will be featured on the exhibit floor throughout the week. ISPOR 2023 has 5 poster sessions containing approximately 400 research posters per session. Within each poster session, there is a one-hour poster discussion period. Presenters are required to be present at their posters during their assigned discussion period.

Poster Tours (By Invitation Only)

We are excited to continue the ISPOR poster tour program. Each tour will feature high impact abstracts within a specific topical area. Please check your abstract notification email to see if your poster was selected for a poster tour.

The poster tours will be a 45-minute experience comprised of

- a designated poster tour area within the Exhibit Hall;
- one Tour Guide (poster tour host);
- and selected posters (up to six).

During the Poster Tours, each poster presenter (one author per poster) will be asked to provide a brief overview of their poster (3 minutes). After each poster overview, there will be an interactive discussion (about 5 minutes) between the poster presenter, attendees, and the tour guide before moving to the next poster.

In addition to the designated poster tour time, these posters will be available for viewing during the poster session.

Note: ISPOR was established in 1995 by a small group of dedicated volunteers and visionaries with the goal of serving as a catalyst to advance the science and practice of health economics and outcomes research (HEOR) around the world. This instruction is taken from the website of https://www.ispor.org/docs/default-source/intl2023/poster-guide.

1. Poster presenters are not necessarily present according to the time schedule. (　　)

2. Each presenter will know whether his or her poster was selected for a poster tour through the abstract notification email before the conference. (　　)

3. Poster presenters have no chance to provide a brief overview of their poster. (　　)

4. There will be an interactive discussion between the poster presenter, attendees, and the tour guide during the post tour. (　　)

Task 10.6　Here is the dialogue between the poster presenter and the attendee at the poster session. Scan the QR code to access the poster and complete the missing information according to the poster below.

A: Excuse me.

B: Yes?

A: May I introduce myself? I am Dr. Green from California State University.

B: How do you do? What can I do for you?

A: Well, I'd like to ask a couple of questions about your poster.

B: All right.

A: Why did you conduct this research?

B: ____.

A: What did you mainly find?

B: ____.

A: Thank you very much. Now I understand your position much better.

B: Oh, it was a pleasure. I always enjoy meeting people who are interested in my research.

A: Bye!

B: Bye!

Delivering a paper presentation

We have previously discussed how to write a research paper, and by now, you should have a clear understanding of what information should be included in each section. However, it's important to note that there are differences between the spoken and written forms of communication. Simply bringing a written paper and reading it word by word at a conference may not be effective. Written language is characterized by long, complex sentences, formal vocabulary, and an impersonal style, whereas spoken language tends to consist of short, simple sentences, informal words, and a personal style. When giving a conference speech, it is necessary to adjust the language style accordingly.

Some people are comfortable giving a speech with just an outline, simple notes, or PowerPoint slides. Others may prefer to write out their entire speech in advance and refer to detailed speaking notes during the delivery. In any case, having a speech script can be helpful.

An oral presentation typically consists of three parts: introduction, body, and conclusion (see Table 10.1). A successful presentation gets straight to the point, captures the audience's attention from the beginning, convinces them with appropriate evidence, and leaves a strong impression with a powerful ending.

Table 10.1 Framework of a paper presentation

Paragraph	What it is about
Introduction	Beginning with something catchy
	Highlighting the motivation of the research
	Establishing the speaker's credibility
	Stating the aim of the study
Body	Stating the problem and research background
	Describing the major pieces of equipment used and recapping the essential step of what was done
	Presenting the data or the end product of the study, test, or project
	Explaining what the results show, comparing results with theory, and evaluating limitations
Conclusion	Making a summary
	Presenting the ending
	Inviting questions

Due to the time constraints inherent in a standard conference presentation, it is important to convey the most important information concisely. Again, if you have already prepared a manuscript for your work, it will simplify the process of preparing your talk. Usually, we would estimate three slides per minute to be an appropriate pace for a scientific presentation. Table 10.2 shows the approximate number of slides for each part and the contents to present in the slides.

Table 10.2 Structure of presentation slides

Section	Slides	Content
Title slide	1	Title, authors, and affiliations
Declaration	1	Possible conflicts of interest Often prescribed format
Introduction	2–3	Background and purpose of study Statement of aims and hypothesis
Materials and methods	3+	Summary of investigative method Illustrations useful
Results	2–4	Short presentation of relevant result data Usually includes tables/graphs
Discussion	2–4	How does your work fit with previous work? What are the implications of the results? Does your work answer questions raised by others? Does your work raise new questions?
Conclusion	1–2	Clear statement of conclusions

Task 10.7 Below is the introduction of a conference speech delivered by Dennis Francis, President-elect of the seventy-eighth session of the United Nations General Assembly who addresses the 74th plenary meeting of the General Assembly. Read it carefully and answer the questions following the text.

Mr. President of the General Assembly,
Mr. Secretary-General,
Colleague Permanent Representatives,
Excellencies,
Delegates,

Allow me, Mr. President, to congratulate you on your astute leadership and management of the General Assembly during the current session. Your calm demeanor and steady hands confer on our deliberations an aura of assurance and control which usually profits multilateral processes.

Today, I stand humbly before this august house of Plenipotentiaries, with immense gratitude to all 193 Member States for the confidence they have reposed in me as I prepare to serve as President of the seventy-eighth session of the General Assembly of the United Nations, an undertaking that is both an honour and a privilege. I have been overwhelmed and yet, at the same time, buoyed by the extraordinary demonstrations of support, solidarity and goodwill that have consistently accompanied me on this journey over recent months. My heart is truly full, even as I remain keenly aware that being called to serve as President of the Assembly entails a weighty responsibility.

It is often said that education is the great liberator, lifting people up the social and economic ladder and strengthening society in the process. Not only is that assessment valid. it is also a truism. I exemplify that pattern, having over the course of my career repeatedly found myself in places where I was called upon to assume onerous responsibilities.

Such experiences would never have materialized had I not had great parents who appreciated the extraordinary power and potential of education and, further, had I not been part of a generation that benefited from an enlightened government policy which challenged and democratized the colonial practice whereby an education was, under State policy, reserved exclusively for the privileged, that is, only those with the means to acquire it.

Therefore, when, in the context of the Sustainable Development Goals, we postpone or neglect to offer support to millions the world over who lack access to quality education, aren't we consigning them facelessly to an intergenerational cycle of poverty, degradation and misery, from which they are hardly likely to break free? It seems to me that the more pragmatic choice would be to do all that we can materially to save those children and young people from near certain defeat—a defeat of circumstance—by affording them, through education, the option of choice and thus the capacity to self-actualize, to their own benefit and that of their communities and societies.

1. Why does the presenter begin with acknowledgments?

2. What is the function of giving thanks to his parents?

3. What is the problem the speaker is talking about?

4. Which should be used more frequently in a conference speech, active or passive voice?

5. What other features of informal language could you identify in the presentation?

Task 10.8 Below is the conclusion of a conference speech delivered by Dennis Francis. Read it carefully and answer the questions following the text.

I am ever conscious of the sensitivity and weight of responsibility that serving as President of this General Assembly imposes on the incumbent. In this regard, I commit to discharge the responsibilities of the office with transparency, accountability, vigour and dedication, bearing in mind that all members have the same rights.

Upon the admission of Trinidad and Tobago to membership in the United Nations in 1962, Sir Ellis Clarke, our first Permanent Representative, in making a comparison between our population and that of the international community, affirmed that "[w]e have, however, developed in our society tolerance, camaraderie, respect for the rights of others, an unswerving opposition to oppression, injustice and racial discrimination, a love of liberty, a supreme faith in the dignity and worth of the human person, and belief in the value of co-operation." These principles will form the basis of my actions as President of the General Assembly.

I will prioritize encouraging and facilitating meaningful dialogue, in various formats, in order to ensure that there is clarity of priorities and the strengthening of common purpose in the interest of coherence. It is my hope to bring forward, with your help and support, a renewed atmosphere of conciliation, cooperation and shared commitment in addressing the many challenges and seizing every opportunity, however inchoate, before the General Assembly. I will seek to enhance current approaches and adopt new ones with feasible solutions, as we endeavour to deliver, or at least to strengthen the bases for delivering, Peace, Prosperity, Progress and Sustainability.

I count on your support during the impending session and call for your fulsome engagement, in good faith, as we purposively accelerate our action towards the achievement of sustainable development for the benefit of all.

Permit me to end on a personal note by expressing my special thanks to my hardworking staff at the Mission whose exceptional devotion to duty and professionalism is a matter of great personal pride and satisfaction.

I should also like to thank my six siblings for their unstinting love and loyalty, and who, I know, will say to me, "Don't think you're the President here—you are still the last." I nevertheless thank them for a lifetime of support and for being here with me today, either in person or virtually, to share in this moment of jubilation. I thank also my lifelong friends from the Class of 1973 of Woodbrook Secondary School, whose genuine friendship and love I can never get too much of. And finally, my grateful thanks to my dear wife Joy, whose smile lights up my day, every day, and whose love, support and encouragement contributed in no small way to making this day possible.

Finally, I share with you a recent discovery: the Latin translation of the words "All Glory to God" is Soli Deo Gloria—SDG. Perhaps more than a coincidence.

Thank you, Mr. President.

Paragraph	Fulfilled or not?	Lexical markers
To signal the end		
To restate the main points		
To mention the future work		
To end the presentation with acknowledgements		
To invite questions		

Other forms of conference communication

After delivering an oral presentation, presenters are typically faced with question and answer (Q&A) sessions, which require impromptu speaking skills. If you know the answer, respond concisely and directly. If you don't know the answer, admit it honestly and skillfully, promise to provide a response later, or seek assistance from others. The key to a successful Q&A session is to have a deep understanding of your topic and be well-prepared, particularly for common questions about research basis and methods.

Apart from Q&A sessions, presenters have opportunities to engage with other participants in various settings such as the reception desk, coffee breaks, and banquets. These interactions provide a chance to exchange thoughts on speeches, research studies, and even travel plans after the conference. Engaging in conversations with professionals both within and outside the meeting can enhance your knowledge of current developments in a specific research area and facilitate learning from others. The primary purposes of such interactions are to exchange information, foster mutual understanding, explore potential collaborations, and establish professional connections. Developing effective communication skills with fellow experts will enrich your experience at academic conferences.

Task 10.9 Some conventional questions at a conference Q&A session have been listed below. Please go through all the questions and work in pairs to ask and answer these questions.

1. Why did you carry out this research? What gap were you trying to fill?
2. Are there any other research groups working in this area? If so, are their findings similar to yours?
3. What key papers did you read while preparing your research?
4. What did you enjoy most about doing your research?
5. What do you think your key finding was?
6. In your opinion, what are the limitations to your research?
7. Could you repeat your main conclusion?
8. Have you published a paper on this topic?
9. What research are you planning for the future?
10. We are doing similar research. Would it be possible for us to see your full results?

Task 10.10 There are three dialogues recorded at an academic conference. Role-play the conversation first and then match the dialogue with the possible situation it may occur.

Dialogue 1

A: Have you registered yet?

B: Not yet.

A: Put your signature here? We need to check the attendance list.

B: By the way, I haven't paid my registration fee yet.

A: You don't need to pay for the registration.

B: Where should I collect my badge?

A: Here is your badge. You are requested to wear it all the time.

Dialogue 2

A: I am a graduate student from XXX University. I have learned about your research on your homepage. Really impressed by the brilliant ideas in your recent publication.

B: Thank you very much. You mean the article I published this May?

A: Yes. The title is "Theories and Methods of Landscape Architecture Heritage Protection."

B: Currently I am writing a paper concerning the latest development in landscape architecture heritage protection.

A: Could you send me a copy of your paper when it is published?

B: I would be happy to. May I have your e-mail address?

Dialogue 3

A: Hello, how did your session go?

B: It was very informative and I wrote down six pages of notes. Can you imagine that?

A: I did not get as much benefit as I had expected. One speaker had a strong accent in speaking English.

B: That's the problem many speakers have. In my session, some speakers used jargon, which was Greek to me. If they could explain them a little bit in the beginning, this would be much better.

A: Yet, I've still got lots of inspiration from the speech.

B: Yes. We don't need to catch everything from the session.

A: Oh, let's have some food. What's your favorite dish?

B: Okay, I prefer vegetables and fruits.

Dialogue 1	At the coffee break
Dialogue 2	At the banquet
Dialogue 3	At the registration

Task 10.11 Scan the QR code to access a speech given by the winner of the Five-Minute Research Presentation. Please watch it carefully and answer the following questions.

1. How does the speech establish its logical connections?

2. What are the signposts and transitions employed in the speech?

3. How does the speaker employ them effectively?

Check your understanding

Task 10.12 Imagine you are going to deliver an oral presentation at an international conference in your research field. Design a 10-page slide for your presentation using PowerPoint. Decide what to put in your first and last slide. Choose a proper title for each of the rest slides and make sure to remove all the redundancy. Present it to your group and discuss with your teammates whether there are any possible improvements.

Unit task

Research Presentation Competition

Up to now, you have already learned how to deliver your presentation both visually and orally. You have tried to make posters and slides, poster presentations and paper presentations. It is time to set about presenting your own research or research proposal in a college students' competition named "Five Minute Research Presentation." Do the following to finish this task.

Step 1: Read the guideline and competition requirements on the website http://sentbase.com/cn5mrp/.

Step 2: Prepare for the competition through the following steps:

- Submit the abstract of the paper.
- Make PPT slides.
- Write a script for your oral presentation.
- Submit a 5-minute video of your research presentation.

Step 3: Study the following rubrics for research presentations carefully, and evaluate your performance.

Content (40%)	1. Clarity: Illustrate the design, background, purpose and significance of the speech 2. Integrity: Fulfill the completeness of the speech with an IBC (Introduction–Body-Conclusion) structure 3. Appropriateness: Explain concepts or difficult points with information or examples that the audience can understand 4. Logic: Justify the major points of the speech logically
Delivery (30%)	1.Verbal skills: Maintain the attention with the audience by pronunciation and intonation 2. Non-verbal interaction: Interact with the audience using body language, eye contact and other non-verbal communication methods properly 3.Visual aids: Use PPT, audio and video clips, pictures and other multimedia aids effectively to facilitate the speech delivery
Language (30%)	1. Accuracy: Use appropriate words and sentence patterns to achieve a certain degree of grammatical accuracy 2. Fluency: Use cohesive devices to deliver the speech coherently and clearly 3. Complexity: Achieve lexical variety and syntactic complexity to some extent

Bibliographies

Aristotle. *On rhetoric: A theory of civic discourse*. 2nd ed. Trans. George A. Kennedy. New York: Oxford UP, 2007.

Sinclair R. Mood, categorization breadth, and performance appraisal: The effects of order of information acquisition and affective state on halo, accuracy, information retrieval, and evaluations. *Organizational Behavior and Human Decision Processes*, 1988, 42: 22–46.

Wallace R J, Czerkawski J W, & Breckenridge G. Effect of monensin on the fermentation of basal rations in the Rumen Simulation Technique (Rusitec). *Br J Nutr*, 1981, 46(1):131–148.

Sample Articles (SA1-SA6):

1. Zhang, X., & Lu, X. Testing the relationship of linguistic complexity to second language learners' comparative judgment on text difficulty. *Language Learning*, 2024, 74(3), 672–706.

2. Silver, D. A theoretical framework for studying teachers' curriculum supplementation. *Review of Educational Research*, 2022, 92(3), 455–489.

3. Tate, T., & Warschauer, M. Equity in online learning. *Educational Psychologist*, 2022, 57(3), 192–206.

4. Gómez, M., & Rodríguez, J. C. Analysis of the convergence of environmental sustainability and its main determinants: The case of the Americas (1990–2022). *Sustainability*, 2024, 16(16), 1-21.

5. Xuan, M. L., Ngoc, L. K., & Thao, L. T. Factors hindering student participation in english-speaking classes: Student and lecturer perceptions. *SAGE Open*, 2024, 14(3), 1–18.

6. Wang, Y., & Feng, D. History, modernity, and city branding in China: A multimodal critical discourse analysis of Xi'an's promotional videos on social media. *Social Semiotics*, 2023, 33(2), 402–425.